Navigation: Science and Technology

Volume 15

This series *Navigation: Science and Technology (NST)* presents new developments and advances in various aspects of navigation - from land navigation, marine navigation, aeronautic navigation to space navigation; and from basic theories, mechanisms, to modern techniques. It publishes monographs, edited volumes, lecture notes and professional books on topics relevant to navigation - quickly, up to date and with a high quality. A special focus of the series is the technologies of the Global Navigation Satellite Systems (GNSSs), as well as the latest progress made in the existing systems (GPS, BDS, Galileo, GLONASS, etc.). To help readers keep abreast of the latest advances in the field, the key topics in NST include but are not limited to:

- Satellite Navigation Signal Systems
- GNSS Navigation Applications
- Position Determination
- Navigational instrument
- Atomic Clock Technique and Time-Frequency System
- X-ray pulsar-based navigation and timing
- Test and Evaluation
- User Terminal Technology
- Navigation in Space
- New theories and technologies of navigation
- Policies and Standards

This book series is indexed in **SCOPUS** and **EI Compendex** databases.

Donghua Zhao

Intelligent Information Processing for Polarization Compass and Inertial Integrated Navigation System

中国工信出版集团 電子工業出版社 PUBLISHING HOUSE OF ELECTRONICS INDUSTRY http://www.phei.com.cn

Springer

Donghua Zhao
Tianjin Renai College
Tianjin, China

ISSN 2522-0454 ISSN 2522-0462 (electronic)
Navigation: Science and Technology
ISBN 978-981-97-7134-9 ISBN 978-981-97-7135-6 (eBook)
https://doi.org/10.1007/978-981-97-7135-6

Jointly published with Publishing House of Electronics Industry
The print edition is not for sale in China (Mainland). Customers from China (Mainland) please order the print book from: Publishing House of Electronics Industry.

This Springer imprint is published by the registered company Springer Nature Singapore Pte Ltd.
The registered company address is: 152 Beach Road, #21-01/04 Gateway East, Singapore 189721, Singapore

Preface

Bioinspired polarized light navigation, as a new type of highly autonomous navigation manner, is based on the principle of insects using atmospheric polarization pattern for autonomous orientation. This orientation method possesses the advantages of strong autonomy, good concealment, large working range, and long working time. Not only the orientation error is not accumulated over time, but also the orientation signal comes from nature and not affected by human factors in this way. This method has significant guiding significance for solving the scientific problem of satellite navigation information complete rejection together with the accumulation of errors over time when inertial navigation systems work alone, which has become an important direction for the navigation technology development. Accordingly, it has been highly valued by domestic and foreign researchers, and related research work has received strong support from governments, enterprises, and other levels at home and abroad. During the operation of a bioinspired polarization compass, there exists error in its output information due to the working principle and environment of the device itself, which includes the various errors caused by polarization angle image noise, heading angle data noise, and changes in the attitude angle change of a bioinspired polarization compass. In complex environmental conditions that encounter occlusion effects such as clouds, tunnels, and buildings, the bioinspired polarization compass becomes temporarily unavailable, which seriously reduces its orientation accuracy and robustness. Therefore, it is extremely necessary to study the information processing technology for a bioinspired polarization compass. In the process of writing the book, the technical accumulation in the information processing of a bioinspired polarization compass over the years is summarized and is analyzed in detail. These methods mainly include: analyzing and processing the noise in polarization angle images and heading angle data, modeling and compensating for the orientation error, and a seamless combination orientation method based on a bioinspired polarization compass. A detailed introduction to heading angle measurement methods based on the atmospheric polarization pattern is also provided.

In the process of writing the book, I referred to the relevant materials of some brother universities and received guidance and help from many professors in the

Tianjin University and North University of China. I would like to express my gratitude.

Owing to the limited level of the author and the shortcomings in the book, we kindly request criticism and correction from readers.

Tianjin, China
December 2023

Donghua Zhao

Contents

Chapter 1
Introduction

1.1 Development Background and Research Significance

Compared with modern orientation methods such as inertial and satellite navigation, the bioinspired polarization orientation is based on the principle of insect navigation. Not only does it possess strong autonomy, good concealment, large working range, and long working time, but its error also does not accumulate over time. This book takes the application background as six rotor small unmanned aerial vehicles (UAVs) performing low altitude unmanned missions (Fig. 1.1), and carries out research on the challenges faced by the autonomous orientation technology for small UAVs. Although the directional signals originate from natural polarization and are generally not affected by modern information warfare, the bioinspired polariztion compass is temporarily unavailable under complex environmental conditions such as encountering obstacles from clouds, tunnels, and buildings. Accordingly, the integrated orientation technique of the bioinspired polarization compass and inertial navigation system displays great significances in exploiting the natural polarization orientation principle and multi-source information fusion algorithms for fulfillong the autonomous orientation requirements of small UAVs and promoting their combat capabilities.

As a new type of highly autonomous navigation method, the integrated orientation of the bioinspired polarization compass and inertial navigation system has turned into an important direction for the development of navigation technology owing to the great guiding significance for solving the scientific problems of autonomous navigation in complex and unfamiliar environments. The remainder of this book is organized as follows. The orientation method and system for atmospheric polarization pattern is presented in Chap. 2. Chapter 3 describes how to process the noise form the bioinspired polarization compass in detailed. The orientation error modeling and compensation technology for the bioinspired polarization compass is illustrated explictly in Chap. 4. Chapter 5 focuses on the seamless integrated orientation method

D. Zhao, *Intelligent Information Processing for Polarization Compass and Inertial Integrated Navigation System*, Navigation: Science and Technology 15,
https://doi.org/10.1007/978-981-97-7135-6_1

Fig. 1.1 UAV performs low-altitude unmanned missions

and system for bioinspired polarization compass and inertial navigation. Finally, the research in this book is summed up and looked into the future in Chap. 6.

1.2 Bioinspired Polarization Orientation Method

There are many organisms in nature that can use the atmospheric polarization pattern for orientation, such as bees, sand ants, dragonflies, and so on. These living things are abale to return to their nests hundreds of meters away by the way of a nearly straight path after a complex and tortuous foraging process in the absence of a significant reference. The compound eye of the insects is the main visual organ as shown in Fig. 1.2. The dorsal edge region (DRA) is a small area facing the sky in which the small eyes are highly sensitive to the distributed pattern of atmospheric polarization.

With regard to the research on the distributed characteristics and characterized methods of the atmospheric polarization pattern, the relevant research work was carried out abroad as early as the sixteenth century. In 1669, Danish physicist B. Erasmus discovered the polarization phenomenon of light for the first time in an experiment, which opened up a new field for light research [1]. Thereafter, French scholar J. Babinet, K. Coulson, British physicist D. Brewster et al. have carried out a large number of in-depth experiments on the causes and characterized methods of atmospheric polarization to further develop and improve the basic theory of atmospheric polarization [2]. In 1871, British scientist Rayleigh deeply studied the

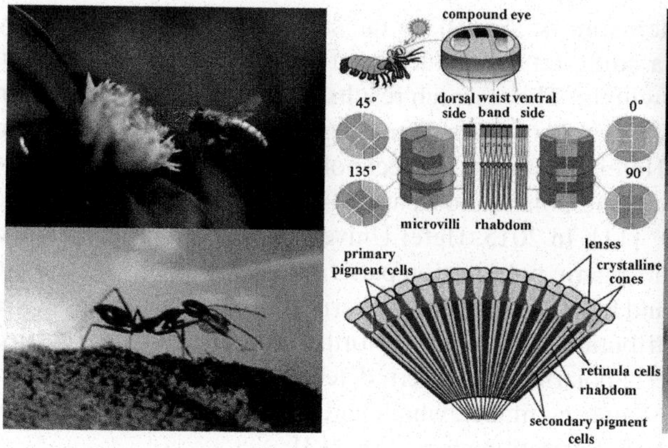

Fig. 1.2 Insect compound eye structure and bionic compound eye model

scattering effects of small particles in the air on light and proposed the classical Rayleigh scattering theory. Rayleigh scattering theory provided an important theoretical support for the later study of the distribution pattern of atmospheric polarization in clear sky on the premise of ignoring the multiple scattering of complex particles [3]. In 1908, German physicist G. Mie proposed a multi-scattering model of light based on spherical particles, namely the Mie scattering theory. The Mie scattering theory pushed the study on single scattering of atmospheric polarization to a multi-scattering study that is more similar to a real scattering model [4]. In 1982, Professor Wehner of the University of Zurich in Switzerland presented the Rayleigh model to describe the distribution pattern of atmospheric polarization in clear weather based on Rayleigh's single scattering theory [5]. In 2017, University of Amsterdam in the Netherlands proposed a three-dimensional Monte Carlo radiative transfer program (ARTES) for simulating scattered light from planetary atmospheres. This transfer program was used to study how three-dimensional atmospheric asymmetry affects polarization signals and evaluate the application of infrared polarization method in direct imaging to observe planetary masses [6]. In 2021, California Institute of Technology (Caltech) used polarization spectroscopy and O-2 near-infrared radiometry to measure the information content of aerosol profiles and evaluate the uncertainty. The results showed that the addition of hyperspectral resolution radiometry and polarization measurement not only reduced the number of required viewing angles, but also effectively improved the polarization and radiation signal-to-noise ratio and detection sensitivity [7].

The research on the distribution and characterization of atmospheric polarization by domestic scholars started late. In 2011, Hefei University of Technology characterized and analyzed the atmospheric polarization pattern based on the complex spherical map and E-vector description. They modeled and simulated the atmospheric polarization pattern under microscopic Rayleigh scattering and proposed an analytical model of the polarization pattern in the sky under the influence of multiple scattering factors [8]. In 2013, Tsinghua University designed an automatic detection device for polarization patterns in the sky [9]. In the same year, Dalian University of Technology conducted a study on the influence of sunlight and moonlight on the polarization pattern of twilight. The research results proved that the polarization pattern of sunlight plays a major role in when formating the sky polarization pattern at twilight [10]. In 2014, Hefei University of Technology proposed a full-Stokes vector-Stokes vector atmospheric polarization pattern measurement system with a field of view of nearly 180° [11]. In 2015, Hefei University of Technology studied a model to simulate the atmospheric polarization pattern Stokes vector based on the Rayleigh scattering model and found that the atmospheric polarization pattern Stokes vector has a "+" shape distribution [12]. In 2015, North University of China carried out a research on atmospheric polarization pattern detection and model reconstruction based on the Rayleigh scattering model, which laid a foundation for subsequent application and algorithm for polarization navigation [13].

In 2018, Hefei University of Technology analyzed the macroscopic variation law of the atmospheric polarization pattern and clarified the modeling method for the "∞" character characteristics of the atmospheric polarization pattern, which improved the feasibility of the course information for polarization navigation [14]. Based on the relative relationship between the oblique attitude coordinate system and the horizontal coordinate system, a modeling method for the atmospheric polarization pattern in the oblique attitude was proposed [15]. In 2021, the Key Laboratory of the National Meteorological Satellite Center used remote sensing multi-angle technology to combine the polarization and anisotropy of atmospheric reflection with the observations of lidar and directional polarization cameras, respectively. They analyzed and studied the aerosol characteristics as multi-angle pure intensity signals and multi-angle polarization signals [16]. These theoretical studies provide a basis for the application of atmospheric polarization pattern.

The above research status shows that the distributed characteristics and characterized methods for atmospheric polarization light have received extensive attention from researchers at home and abroad. Meanwhile, atmospheric polarization pattern detection devices have also been applied. The distributed characteristics and characterized methods for accurate atmospheric polarization in complex environments have become a research hotspot, which has opened the prelude to the research on autonomous navigation using sky polarization information. Some canonical characteristics and characterized methods for atmospheric polarization are shown as Fig. 1.3.

In 1982 and 1989, University of Zurich in Switzerland studied how to derive bee celestial maps and special parts of the retina from the E-vector model in the sky, which is shown as Fig. 1.4. They simulated the E-vector directional distribution in the sky to use the polarization pattern as a compass for navigation [17, 18]. Subsequently, some scientists have done a lot of research on the visual perception, flight guidance, navigation, learning and memory of various organisms in nature using their visual and brain polarization-sensitive neurons in complex and unfamiliar environmental conditions such as day/night and mist/clouds. In 2001, University of Zurich in Switzerland discovered that polarization-sensitive neurons (POL neurons) in the cricket visual system can integrate large-area information in the sky. These POL

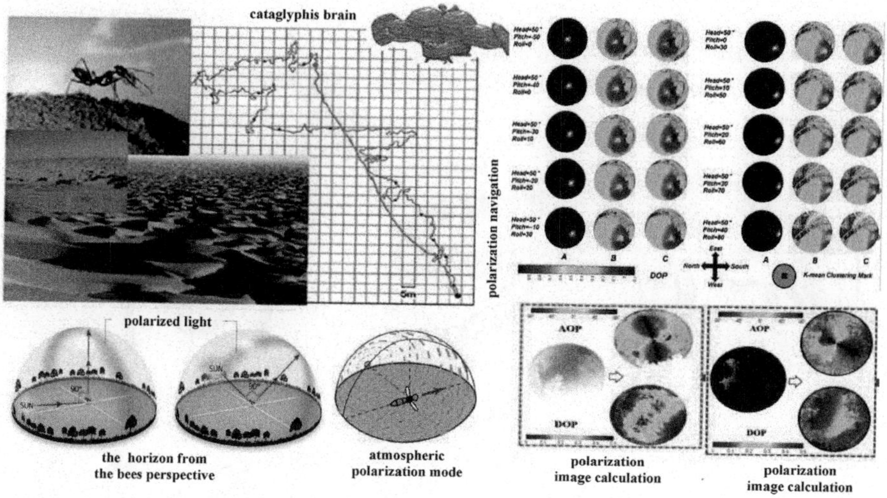

Fig. 1.3 Characteristics and characterized methods for atmospheric polarization

neurons can improve the quality of atmospheric polarization signals and enhance the sensitivity of navigation information by filtering out local perturbations of polarization patterns caused by clouds [19]. In 2011, University of Marburg further studied the response on two descending neurons and the anterior thoracic ganglion of the locust brain to find that the mechanism is able to be emplyed to navigate with the atmospheric polarization E-vector pattern [20]. Figure 1.4 displays honeybee POL neurons to navigate utilizing atmospheric polerization pattern.

In 2015, University of Bristol in the United Kingdom carried out an research hpw to characterize the polarization sensitivity for human eyes. They developed a new method for measuring the lower limit of human perceived polarization light detection by gratings with only polarization contrast at different degrees of polarization [21]. In 2019, Institute of Biological Neuroscience in Berlin conducted a research on the heading selection mechanism of flying Drosophila under different polarization angles and found that these behavioral responses are wavelength-specific. That is, they can be selected under polarized ultraviolet stimulus, but not under polarized green light, which provides further evidence for Drosophila to use celestial cues for visual navigation and course correction [22]. In 2020, a study by University of Maryland State discovered that mantis shrimp can navigate based on the hierarchical path integration of the sun, overhead polarization patterns, and internal unique orientation cues when foraging. This discovery became the first fully underwater animal path integration method discovered so far, which opens up a new way to study the neural basis of navigation behavior and improve navigation methods in arthropods and other diving animals [23]. In 2021, Dron University in Sweden and Simon Fraser University in Canada conducted a detailed analysis that the beetle how to use polarizated light to navigate. This study showed that the beetle used the polarization patterns generated by the Sun, the Milky Way and the Moon to move in a fixed direction during the

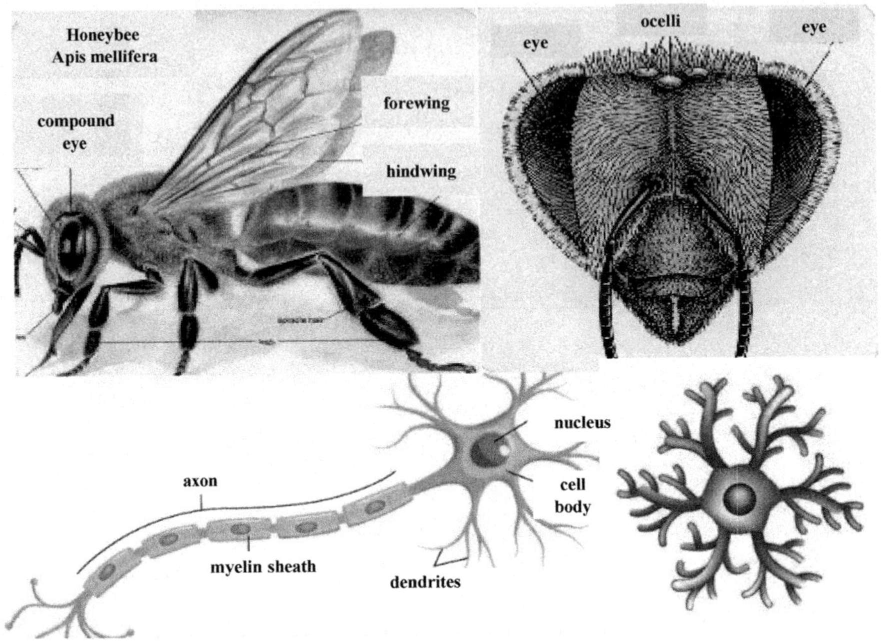

Fig. 1.4 Honeybee POL neurons to navigate utilizing atmospheric polerization pattern

day and night. In the morning and afternoon, the sun was used to determine the direction. At noon, the wind was used to determine the direction and the polarization was mainly relied on to maintain a straight path at night or in the forest [24, 25].

The good application prospect of polarization orientation has also attracted extensive attention from domestic scholars. In 2009, Dalian University of Technology designed a new bionic polarization angle sensor based on the principle of bioinspired polarization orientation they presented an angle error compensation algorithm and carried out test for the navigation sensor prototype by outdoor mobile robots. Subsequently, in 2015, the team proposed a heading angle measurement system based on polarization and MEMS gyroscopes, which can provide accurate heading information for flight control [26, 27]. In 2013, Hefei University of Technology also carried out sensor design research based on the principle of bionics. This research solved the ambiguity problem of polarization navigation angle measurement and proposed a heading angle calculation method based on the symmetry detection method on the basis of atmospheric polarization pattern and the POLneuron model imitating sand ants [28]. In 2014, National University of Defense Technology proposed an angular error calibration method and analyzed the polarization orientation algorithm and error [29]. From 2016 to 2019, the team proposed some bionic navigation algorithms based on the grid cell model/positional cell model and biological navigation mechanism, a multi-eye polarization visual navigation method, a navigation topology map construction method, and a bionic polarization light orientation method based on

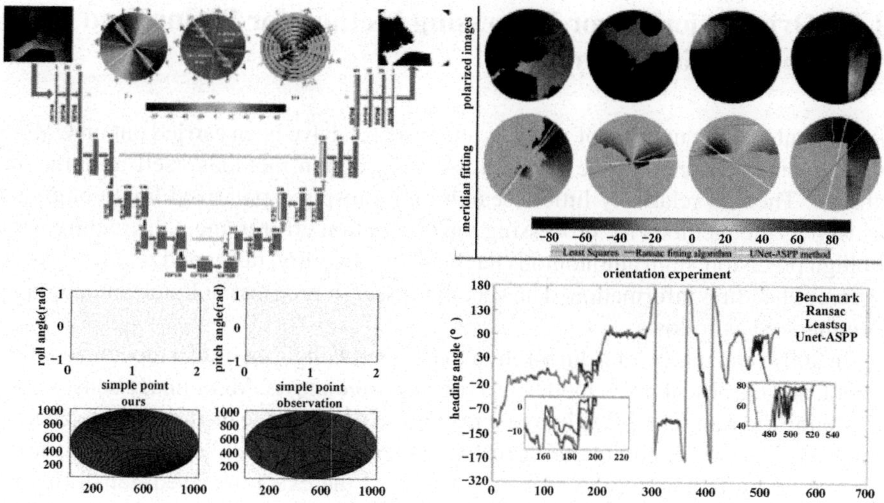

Fig. 1.5 Bioinspired polarization sensor and its navigation test results

the animal hippocampus navigation mechanism and the RatSLAM algorithm as well as for the ground unmanned platform [31–34]. In 2020, the team proposed a model and method for building atmospheric polarization orientation for any pixel based on known information under cloudy weather conditions [35]. In 2015, North University of China proposed a heading angle solution method based on full-sky atmospheric polarization detection and a method for extracting the pitch angle and roll angle of an aircraft based on the stability distribution characteristics of the atmospheric polarization pattern [30]. In 2021, in order to improve the fitting accuracy of the solar meridian and the polarization orientation system accuracy, the team exploited a coarse extraction method for the symmetry axis and then accuratly extracting the solar meridian after continuous rotation [36]. Some classical bioinspired polarization sensor and navigation test results are shown as Fig. 1.5.

It can be seen from the above research at home and abroad that the directional measurement based on bioinspired polarization sensors has achieved good results under laboratory conditions. On the basis of the above research, how to further improve the orientation accuracy in the practical application process has become an important research direction of the bioinspired polarization compass. Therefore, it is necessary to carry out the research on the processing method for the bioinspired polarization compass orientation error.

1.3 Orientation Error Processing Method for Bioinspired Polarization Compass

At present, a large number of analysis and research have been carried out on CMOS/CCD image sensors and the image noise suppression methods itself at home and abroad. There is relatively little research on bioinspired polarized light compasses as well as the methods for processing the orientation error to meet the requirements of high precision, high autonomous navigation capability, faster and real-time calculation of heading information. The specific research progress at home and abroad is summarized as follows.

In 2007, University of Edinburgh in the United Kingdom used a recognized theoretical and empirical noise model to write and simulate a CMOS image sensor noise modeling method for CMOS image sensor noise to verify the method effectiveness in improving image quality in the case of various noise sources [37]. In 2013, Tampere University of Technology in Finland introduced the exact unbiased transformation of the Anscombe transform, which applied a variance stabilization transformation to the noise data. They used the Gaussian denoising algorithm to denoise the stabilized transformed data and finally employed the inverse variance stabilization transformation to the denoised data for reconstruction [38]. In 2016, School of Information Technology at COMSATS proposed a goodness-of-fit (GOF) statistical test method for CMOS image denoising based on discrete wavelet transform (DWT) and empirical distribution function (EDF) statistic, transforming the denoising problem into a hypothesis test problem by using a null hypothesis corresponding to the existence of noise and another hypothesis indicating that only the desired signal exists in the measured image sample [39]. In 2019, University of Consat raised a Poisson noise or Poisson-Gaussian hybrid noise removal method based on detection theory and hypothesis testing integrated with variance stable transformation for CMOS/CCD imaging sensor noise [40]. In 2020, Gdansk University of Science and Technology proposed a digital clock stop technique for gain and offset correction of time-pattern analog-to-digital converters. They solved the problem of fixed-pattern noise suppression and linearity improvement of massively parallel CMOS imagers in time pattern, which is essential for compensation of dark current non-uniformity noise and light response non-uniformity noise [41]. In 2021, Nanyang Technological University designed an approximate array multiplier suitable for digital image denoising by making vertical or horizontal cuts in a precision array multiplier with different input and output allocations [42]. System of CMOS imaging acquisition and image de-noising results are shown as Fig. 1.6.

In 2010, Tianjin University exploited the total least squares method to study the multiplicative interference noise suppression contained in COMS image sensors. This study selected four typical noise reduction algorithms: total least squares method, block matching and three-dimensional filtering, bilateral filtering, and linear filtering to compare the denoising effects [43]. In 2011, the research team conducted a study on the impact of demosaicing noise and different denoising sequences on image quality, which effectively monitored and analyzed the image and noise [44]. In 2017,

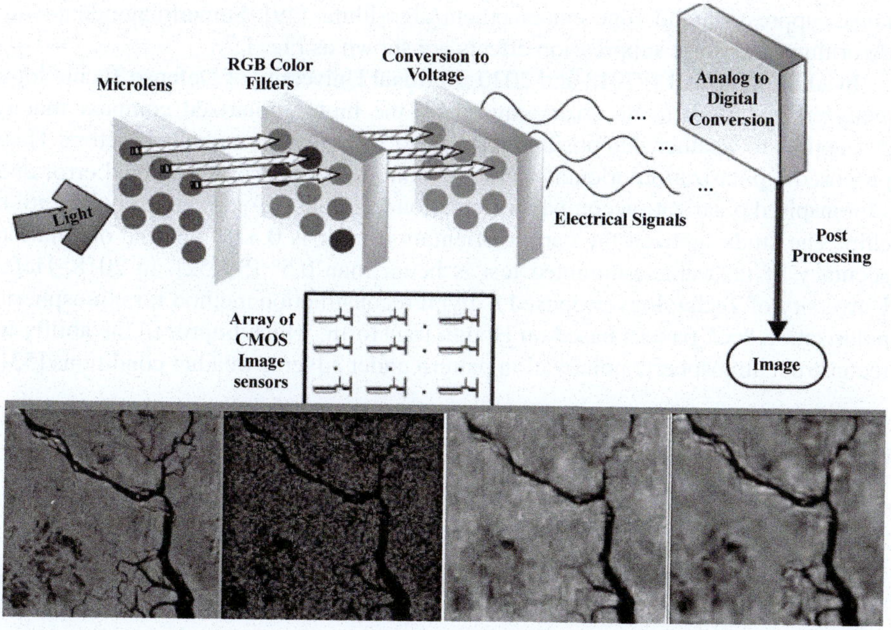

Fig. 1.6 System of CMOS imaging acquisition and image de-noising results

Harbin Engineering University calculated the image noise under different exposure conditions by analyzing the gray value data output by the CMOS image sensor. They corrected the fixed-pattern noise by using the flat-field correction method of least squares fitting and then used the quadratic fitting to solve the fixed-pattern noise nonlinearity problem [45]. In 2019, Hangzhou Dianzi University used the CMOS image sensor signal-dependent noise model to solve the problem of traditional image noise reduction algorithm in suppressing noise while reducing useful signal information starting from signal power and energy. They aimed at the characteristics of signal-dependent noise, that is, different gray values are affected by different intensity noise. Finnaly, they proposed a random resonance image noise reduction algorithm based on image segmentation and a cascaded random resonance system design method [46]. In 2020, Tianjin University proposed a convolutional neural network image denoising method based on a new three-channel loss function and network structure for the noise in the real image from CMOS image sensors, making convolutional neural network more suitable for denoising processing [47]. In 2021, Beijing Institute of Technology proposed a dedeforming kernel prediction neural network joint denoising method for a self-developed enhanced CMOS image sensor [48]. National Taipei University in Taiwan proposed a module based on depth image prior for noise generated in images collected or transmitted in low-light environments. The image fusion module and the progressive refinement module based on the Laplace pyramid decomposition integrated with improved CMOS image sensor noise suppression and edge enhancement algorithms [49]. SomeiImage denoising algorithms and noise suppression effects are shown as Fig. 1.7.

In addition, in 2014, 2018 and 2020, National University of Defense Technology analyzed and studied the errors caused by the bionic polarized compass that is susceptible to weather, sun position, carrier horizontal angle and other factors. They proposed a polarization orientation model containing polarization model error and a bioinspired polarization compass orientation technology for pixelated polarization chip. The study realized the carrier orientation error is 0.323° and the orientation accuracy of the vehicle-mounted test is better than 0.5° [50–52]. In 2018, Hefei University of Technology proposed a glyph reconstruction method for atmospheric polarization "∞" pattern based on Hough transform, which improved the ability to reconstruct atmospheric polarization pattern under adverse weather conditions [53].

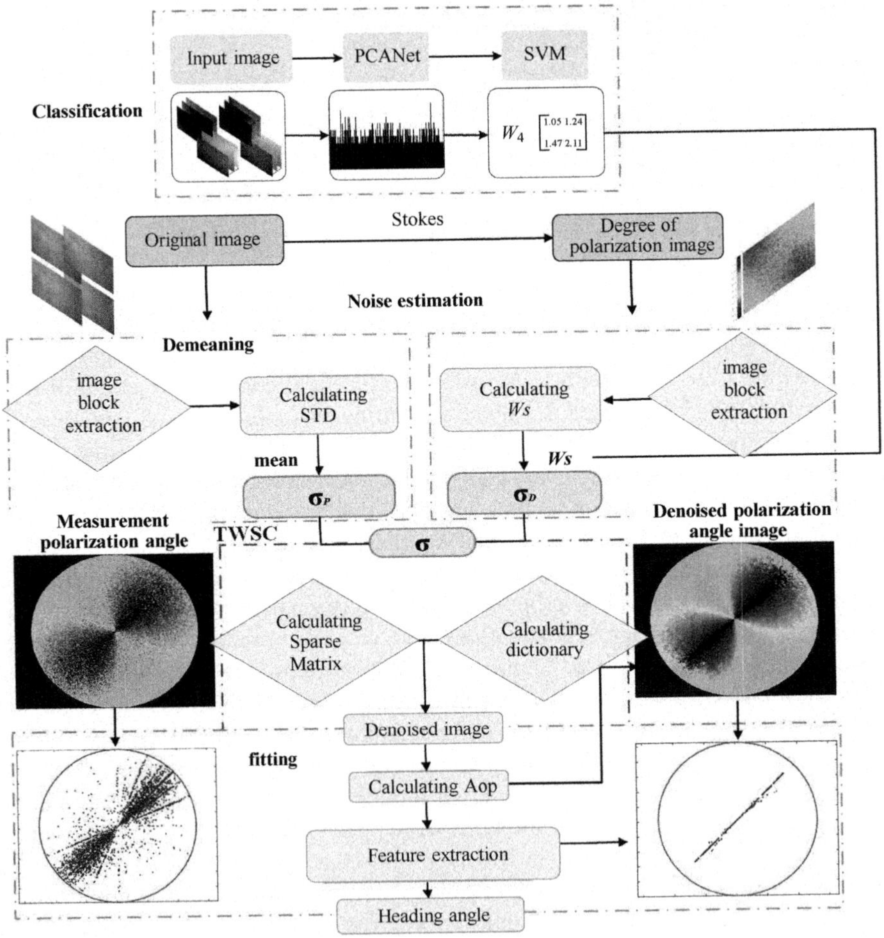

Fig. 1.7 Image denoising algorithms and noise suppression effects

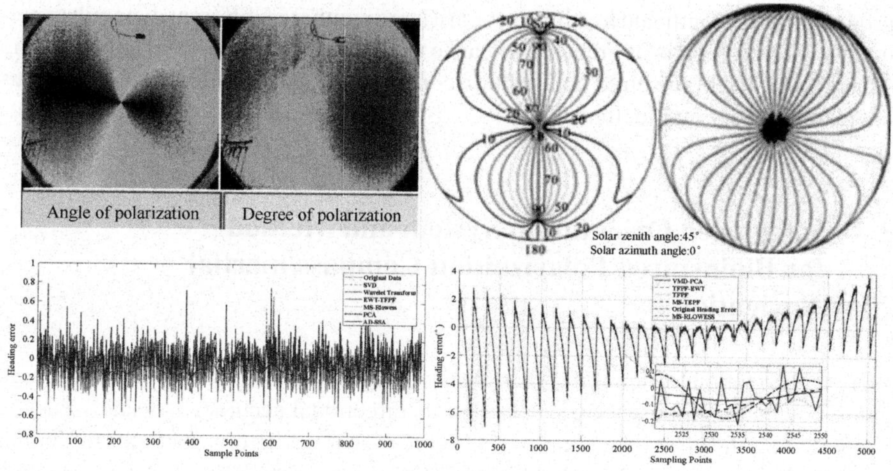

Fig. 1.8 Processing methods and effects for the polarization compass orientation error

In the same year, University of Science and Technology of China conducted a study on the influence of turbid atmosphere on polarization navigation, which effectively improved the orientation accuracy by reducing the error caused by the turbid atmosphere on polarization navigation [54]. In 2020, North University of China proposed a polarization compass heading error processing method based on the integrated of variance stable transform polarization image denoising and inclination error compensation of high-efficiency extreme learning machine. They solved the problem that the noise in the polarization angle image measured by the polarization compass and the inclination angle of the compass during the working process cause the orientation accuracy to be significantly reduced [55]. Some processing methods and effects for the polarization compass orientation error are shown as Fig. 1.8.

From the above research at home and abroad, it can be seen that the noise suppression technology for COMS/CCD image sensors is relatively mature by using different algorithms or image segmentation methods, which provides technical feasibility for the research work of the bioinspired polarization compass denoising in this book. The bioinspired polarization compass studied in this book calculates the heading angle based on polarization images and outputs the heading angle data through FPGA circuits. The noise source and its generation mechanism are different. This noise comes not only from the polarization image acquired by imaging system in the polarization compass, but also from the heading data output by the compass. Therefore, it is necessary to study the noise processing method for high-precision bioinspired polarization compass.

In addition, the bioinspired polarization compass fixed to the moving carrier inevitably produces tilt such as pitch and roll during the actual motion of the carrier. The orientation error caused by these inclination changes cannot be ignored. At the same time, we found that the coupling of the angle between the solar meridian and the axis of the carrier body (A-SMBA) and the inclination angle of the carrier will produce a more significant heading error in a large number of different dynamic experiments. Therefore, it is of great significance to study the orientation error processing strategies for the bioinspired polarization compass in depth to enrich the autonomous navigation theory and method.

1.4 Integrated Orientation System and Method for Bioinspired Polarizaition Compass/inertial Navigation

Inertial Navigation System (INS) has the advantages of strong anti-interference, high autonomy and high output frequency, but its directional accuracy is usually easy to decrease significantly over time. As a bionic visual navigation method, the bioinspired polarization compass (PC) has been widely used in UAV orientation research due to its advantages of high autonomy and durability. However, PC has poor adaptability to the environment and is susceptible to harsh environmental influences such as clouds or tunnels, resulting in a significant decrease in its directional accuracy. Therefore, the PC/INS integrated orientation method can give full play to the advantages of PC and INS. That is, the problem of divergence of INS heading measurement values with time can be corrected through PC. INS can still obtain orientation information and maintain high orientation accuracy when PC is temporarily unavailable due to harsh environment. It is of great significance to improve the robustness and orientation accuracy for the whole integrated orientation system.

In 2006, Lund University in Sweden proposed an absolute (i.e., geographical) direction system by the way of migratory birds to orient themselves at sunrise and sunset. These birds make full use of a integrated compass system of geomagnetic field, star, sun, and polarization to determine the direction of migration and provide a primary calibration reference for all their compasses [56]. In 2010, University of Massachusetts in the United States found that monarch butterflies use time-compensated solar compasses and magnetic compasses to navigate long-distance migration in autumn [57]. In 2013, Kurztown University conducted an in-depth study on the navigation mechanism of honeybees and found that honeybees have at least three kinds of compass devices: a magnetic compass, a solar compass based on the sun daily rotation and a skylight pattern connected to the sun, and a spare celestial compass based on the sun memory of the sun movement over time relative to the terrain [58]. In 2018, Lund University in Sweden conducted an in-depth study on the effects of different solar and magnetic compasses on the migration routes of migratory birds during their spring andautumn migration. The results showed that different compasses were used for different latitudes, migration directions, migration seasons, and geographical locations. They could be successfully reached by reorienting along the migration route once or more times, or using map information [59]. In 2019, University of Marseille proposed a polarization compass and a minimum light flow sensor integrated navigation system based on Michaelis–Menten adaptive

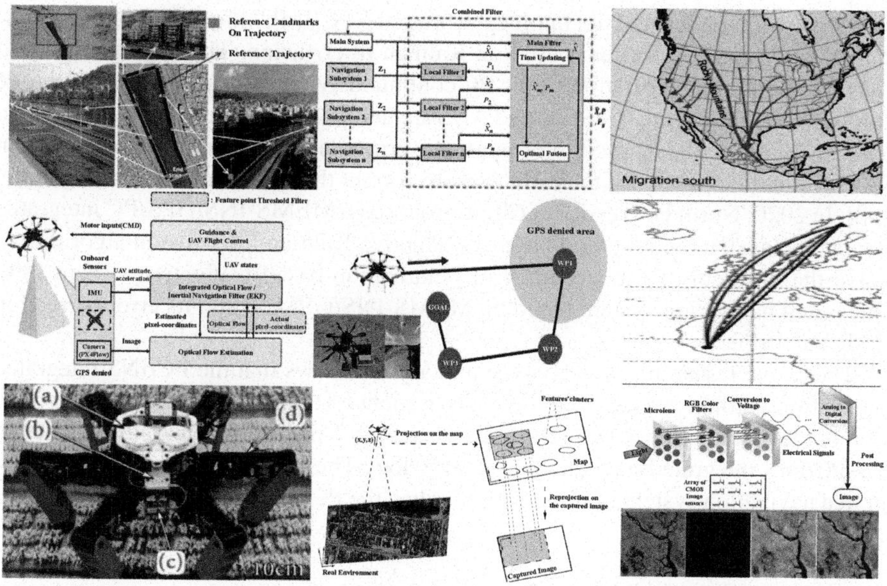

Fig. 1.9 Outdoor autonomous navigation with minimum light flow sensor navigation system

pixels. This integrated navigation system can be applied to the autonomous navigation for outdoor robots. The test results showed that the heading error varies stably between 0.3° and 2.9° under sunny to changeable weather conditions [60]. Outdoor autonomous navigation with minimum light flow sensor navigation system is shown as Fig. 1.9.

In China, from 2007 to 2010, Harbin Institute of Technology proposed a PC/GPS/INS integrated navigation method based on federated Kalman filter fusion algorithm, a PC/geomagnetic/GPS/SINS integrated navigation method based on improved federated Kalman filter multi-sampling fusion algorithm, a PC/GPS/SINS integrated navigation attitude setting algorithm based on federated Kalman filter, and a PC/based distributed Kalman filterLight flow/GPS/INS integrated navigation methods, respectively [61–64]. In 2015, in order to improve the accuracy of UAV attitude measurement, Dalian University of Technology proposed a PC/GPS/SINS integrated navigation method based on centralized Kalman filter algorithm and complementary filter algorithm [65]. In 2016, the National University of Defense Technology proposed a binocular vision/MIMU/PC integrated navigation method based on recursive topology graph nodes, which has the advantages of positioning error not accumulating with time and accuracy not affected by the initial heading angle [66]. In 2017, Nanjing University of Science and Technology proposed a PC/Beidou/INS integrated navigation method to solve the low accuracy problem of microelectromechanical inertial measurement units. They realized the extraction method for carrier three-dimensional attitude information and ensured the advantages of high accuracy and robustness of the integrated navigation system [67]. In 2019, North China

University of Science and Technology proposed a PC/IMU/Air Data System (ADS) integrated navigation system based on Kalman filter fusion. The system measured the heading angle information by the PC continuously outputs the speed and altitude information through the ADS. PC and ADS are integrated to solve the IMU error drift. The Kalman filter is adopted to estimate the error state of the integrated navigation system and then put to use the estimation to correct the systematic error in real time [68]. In 2020, North University of China proposed a MEMS-INS/GPS/PC integrated navigation system and method based on adaptive Kalman filter fusion algorithm to solve the problem of low accuracy of polarization-based navigation system [69]. Integrated navigation system based on MEMS-INS/GPS/PC and its error correction is shown as Fig. 1.10.

The above research status at home and abroad shows that the PC/INS integrated navigation method has realized the measurement of the attitude, velocity and position of the carrier. However, in the previous research results, the output frequency of the PC/INS integrated navigation system is low. The robustness of the whole integrated navigation system is still unsolved when the PC is unavailable due to complex

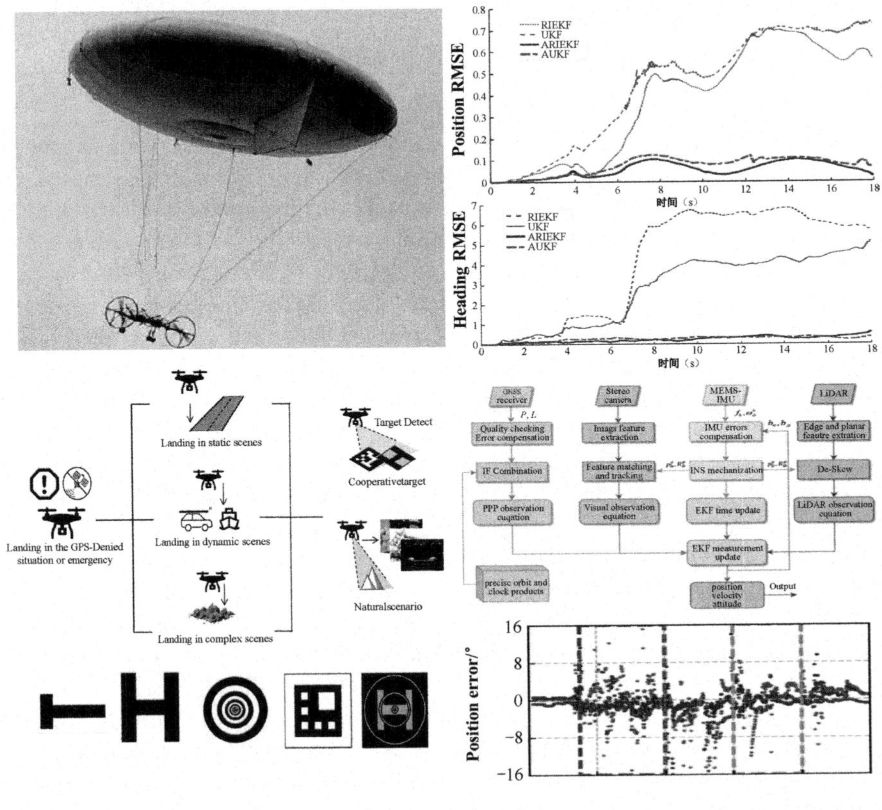

Fig. 1.10 Integrated navigation system based on MEMS-INS/GPS/PC and its error correction

environmental influences. As a result, it is a key scientific problem to be solved urgently how to fuse PC and INS effectively through filtering algorithms to improve the robustness and orientation accuracy for the integrated system.

In summary, the orientation methods and theories based on the bioinspired polarization compass and inertial navigation have become more and more mature since they have shown strong autonomy in experimental verification. On the basis of the existing research results, this book intends to carry out a research on the high-precision bioinspired polarization compass orientation error processing and PC/INS seamless integrated orientation method and system under complex environmental conditions.

References

1. Horváth G, Varjú D. Skylight polarization. Polarized Light Anim Vis Polarizat Patterns Nat. 1669;1(7):18–22.
2. Arago F. Astronomie populaire. Gide et J. Baudry; 1854.
3. Rayleigh L. On the light from the sky, its polarization and color. Phil Mag. 1871;61:107–279.
4. Mie G. Beitrage zur optik triiber medien. Speziell kolloidaler Metallosungen Annalen der Physik. 1908;25:337–444.
5. Wehner R. Himmels navigation bei Insekten. Neurophysiologie. 1982;1(184):132.
6. Stolker T, Min M, et al. Polarized scattered light from self-luminous exoplanets three-dimensional scattering radiative transfer with ARTES. Astron Astrophys. 2017;607:42.
7. Choi M, Sander SP, et al. Aerosol profiling using radiometric and polarimetric spectral measurements in the O-2 near infrared bands: estimation of information content and measurement uncertainties. Remote Sens Environ. 2021;253:112179.
8. Jun G, Song J, Zhao X, Jinjin L. Skylight polarization modeling and its simulation in micro-rayleigh scattering theory. J Syst Simul. 2011;24(3):677–84.
9. Kaichun Z, Hao L, Zheng Y. Automatic detection system for skylight polarized pattern. Opt Precis Eng. 2013;21(2):239–44.
10. Yan C, Qisheng G, Jinkui C, Chen C. Influence of sunlight and moonlight on polarization patterns during twilight. Opt Precis Eng. 2013;21(1):34–9.
11. Zhongshun Z. The design and implementation of full stokes vector image type atmosphere polarization mode measurement system. Hefei University of Technology; 2014
12. Ziqian W, Zhiguo F, Xudong Z, Xin W. Stokes vector modeling and simulation of atmospheric polarization pattern based on Rayleigh scatter. Comput Appl Softw. 2015;32(7):47–50.
13. Chenguang W, Jun T, Jiangtao Y, Jun L. Detection and model reconstruction of atmospheric polarization pattern based on Rayleigh scattering theory. J North Univ China. 2015;36(5):570–5.
14. Zhiguo F, Chao X, Chuan W, Jun G. Atmospheric polarization pattern and its application in autonomous navigation. Remote Sens. 2017;45(3):1–7.
15. Daqian W, Jun G, Yan S, Zhiguo F. Study on modeling method of skylight polarization mode under tilting attitude. Chin J Sci Instrum. 2018;39(5):73–80.
16. Si Y, Lu QF, et al. A review of advances in the retrieval of aerosol properties by remote sensing multi-angle technology. Atmosph Environ. 2021;244:117928.
17. Rossel S, Wehner R. The bee's map of the e-vector pattern in the sky. Proc Natl Acad Sci USA. 1982;79(14):4451–4.
18. Wehner R. Neurobiology of polarization vision. Trends Neurosci. 1989;12(9):353–9.
19. Labhart T, Petzold J, Helbling H. Spatial integration in polarization- sensitive interneurones of crickets: a survey of evidence, mechanisms and benefits. J Exp Biol. 2021;204(14):2423–30.

20. Trager U, Homberg U. Polarization-sensitive descending neurons in the locust: connecting the brain to thoracic Ganglia. J Neurosci. 2011;31(6):2238–47.
21. Temple SE, McGregor JE, et al. Perceiving polarization with the naked eye: characterization of human polarization sensitivity. Proceed Royal Soc B Biol Sci. 2015;25(4):E67–8.
22. Mathejczyk TF, Wernet MF. Heading choices of flying Drosophila under changing angles of polarization. Sci Rep. 2019;9:16773.
23. Patel RN, Cronin TW. Mantis shrimp navigate home using celestial and idiothetic path integration. Curr Biol. 2020;30(11):1981–7.
24. Dacke M, Baird E, et al. How dung beetles steer straight. Annu Rev Entomol. 2021;66:243–55.
25. Blake AJ, Couture S, et al. Approach trajectory and solar position affect host plant attractiveness to the small white butterfly. Vision Res. 2021;186:140–9.
26. Zhao K, Chu J, Wang T, et al. A novel angle algorithm of polarization sensor for navigation. IEEE Trans Instrum Meas. 2009;58(8):2791–5.
27. Wei Z, Jinkui C, Yinlong W. Design of heading angle measurement system based on polarization and MEMS gyroscope. Transd Microsyst Technol. 2015;34(1):104–5.
28. Bo W, Jun G, Zhiguo F, Xin W. An angle processing method based on the POL-neuron model of desert ant. Opto-Electr Eng. 2013;10:28–34.
29. Xian Z, Hu X, Lian J, Zhang L, Cao J, Wang Y, Ma T. A novel angle computation and calibration algorithm of bio-inspired sky-light polarizaton navigation sensor. Sensors. 2014;14:17068–88.
30. Wang C, Tang J, Liu J. Solving the heading angle using atmospheric polarization detection in the whole sky domain. Opto-Electr Eng. 2015;12:60–5.
31. Tao M. Research on position and orientationconstraints based bio-inspired navigation. National University of Defense Technology; 2015.
32. Yujie W. Bionic navigation algorithms based on polarization vision. National University of Defense Technology; 2017.
33. Chen F. Research on topological map basedbio-inspired navigation. National University of Defense Technology; 2018.
34. Jing Z, Jun T, Donghua Z, et al. Place recognition with deep superpixel features for brain-inspired navigation. Rev Sci Instrum. 2021;91(12):125110.
35. Fan Y, Xiaofeng H, et al. Atmospheric polarization orientation method in cloudy weather. Acta Aeronaut ET Astronaut Sin. 2019;41(9):1–7.
36. Chengshuai Z, Xindong W, Donghua Z, Xiaojie L, Chenguang W, Chong S. A bionic polarization orientation system based on improved solar meridian fitting method. Navig Posit Timing. 2021;8(04):68–74.
37. Gow RD, Renshaw D, Findlater K, et al. A comprehensive tool for modeling CMOS image-sensor-noise performance. IEEE Trans Elect Dev. 2007;54(6):1321–9.
38. Mäkitalo M, Foi A. Optimal inversion of the generalized anscombe transformation for poisson-Gaussian noise. IEEE Trans Image Process. 2013;22(1):91–104.
39. Rehman N, Naveed K, Ehsan S, McDonald-Maier K. Multi-scale image de-noising based on goodness of FIT (GOF) tests. Eur Sig Process Confer. 2016;24:1548–52.
40. Naveed K, Ehsan S, et al. A multiscale de-noising framework using detection theory with application to images from CMOS/CCD sensors. Sensors. 2019;19(1):206.
41. Klosowski M, Sun YC. Fixed pattern noise reduction and linearity improvement in time-mode CMOS image sensors. Sensors. 2020;20(20):1–15.
42. Balasubramanian P, Nayar R, Maskell DL. Approximate array multipliers. Electronics. 2021;10(5):1–20.
43. Xuan L. Research on CMOS image sensor noise suppression. Tianjin University; 2010.
44. Jiawei Z, Xiaochen L, Donghua Z, Chenguang W, Chong S, Jun T, Jun L. Self-adaptation feature attention network with muti-step fusion for single image dehazing. Tournal Test Measur Technol. 2022;36(04):347–52.
45. Chenglin M. Research of testing technology for main noises of CMOS image sensor. Harbin Engineering University; 2017.
46. Jianxin C. Signal-dependent noise suppression method for CMOS image sensor based on stochastic resonance theory. Hangzhou Dianzi University; 2019.

47. Yahui P, Xiaochen L, Chong S, Haoqian H, Donghua Z, et al. An improved light flow algorithm based on mask-R-CNN and K-means for velocity calculation. Appl Sci Basel. 2019;9(14):2808.
48. Han ZH, Li L, et al. De-noising and motion artifact removal using deformable kernel prediction neural network for color-intensified CMOS. Sensors. 2021;21(11):3891.
49. Huang SC, Hoang QV, et al. An advanced noise reduction and edge enhancement algorithm. Sensors. 2021;21(16):1–12.
50. Xian Z, Hu X, Lian J, et al. A novel angle computation and calibration algorithm of bio-inspired sky light polarization navigation sensor. Sensors. 2014;14(9):17068–88.
51. Baolong Y, Donghua Z, Xiaojie L, Xindong W, Deli Y, Chenguang W, Chong S. Optimized light flow velocity measurement method based on fuzzy kernel clustering algorithm. Navig Posit Timing. 2021;9(3):1–8.
52. Chen F, Xiaoping H, Xiaofeng H, Junxiang L, Yujie W. Influence of skylight polarization pattern on bionic polarized orientation and corresponding experiments. Opt Precis Eng. 2015;23(9):2429–37.
53. Chuan W, Zhiguo F, Chao X, Yan S. Reconstruction method of atmospheric polarization mode ∞ shape based on Hough transform. Chin J Quant Electr. 2018;35(2):267–71.
54. Yawen O, Pengfei W, Lili W. Influence of turbid atmosphere on polarization navigation. Infrared Laser Eng. 2018;47(3):1–10.
55. Chong S, Xindong W, Donghua Z, et al. Comprehensive heading error processing technique using image de-noising and tilt-induced error compensation for polarization compass. IEEE Access. 2020;8:187222–31.
56. Muheim R, Phillips JB, Akesson S. Polarization cues underlie compass calibration in migratory songbirds. Science. 2006;313(5788):837–9.
57. Reppert SM, Gegear RJ, Merlin C. Navigational mechanisms of migrating monarch butterflies. Trends Neurosci. 2010;33(9):399–405.
58. Dovey KM, Kemfort JR, Towne WF. The depth of the honeybee's backup sun-compass systems. J Exp Biol. 2013;216(11):2129–39.
59. Muheim R, Schmaljohann H, Alerstam T. Feasibility of sun and magnetic compass mechanisms in avian long-distance migration. Mov Ecol. 2018;6(8):1–15.
60. Dupeyroux J, Viollet S, Serres JR. An ant-inspired celestial compass applied to autonomous outdoor robot navigation. Robot Auton Syst. 2019;117:40–55.
61. Hongqian L, Xianlin H, Hang Y. Polarized-light-based navigation in three-dimensional space. Light Techn. 2007;32(3):412–4.
62. Mingming L. Polarized-light/geomagnetism/gps/sinsintegrated navigationalgorithm. Harbin Institute of Technology; 2008.
63. Jing Z, Donghua Z, Chenguang W, Chong S, Jun T, Jun L. Inertial-based brain-like navigation strategy based on scene recognition. Navigat Control. 2020;19(Z1):119–25.
64. Jiaqi L. Polarized-light-aided attitudede termination in integrated navigation systems. Harbin Institute of Technology; 2010.
65. Yang R. The research of fusion algoritlml of polarization iilformation based on fixed-wing UAV. Dalian University of Technology; 2014.
66. Xianjibun S. Algorithms for graph based autonomous navigation. National University of Defense Technology; 2015.
67. Wenjie S. Research on polarization imitated ant/BD/MEMS-IMU independent integrated navigation system. Nanjing University of Science and Technology; 2017.
68. Guo XY, Yang J, et al. An autonomous navigation system integrated with air data and bionic polarization information. Trans Inst Meas Control. 2019;41(13):3679–87.
69. Chong S, Xiong Y, Donghua Z, et al. Multi-rate strong tracking square-root cubature Kalman filter for MEMS-INS/GPS/polarization compass integrated navigation system. Mech Syst Sig Process. 2021;163:108146.

Chapter 2
Orientation Method and System for Atmospheric Polarization Pattern

Biological research results demonstrate that insects such as sand ants and bees not only possess the ability to be highly sensitive to polarization information, but also can determine their own movement direction through the distribution of polarization patterns in the sky. The highly autonomous sky polarization pattern is one of the inherent resources in nature, which is basically not damaged and interfered by human subjective factors. It is very suitable to provide orientation information for carriers in the environment of weak or no satellite signals. Consiquently, the orientation method based on the atmospheric polarization pattern has a good application prospect for autonomous orientation. This chapter focuses on the vector-oriented methods and systems based on atmospheric polarization patterns.

2.1 Orientation Method for Atmospheric Polarization Pattern

2.1.1 Analysis and Automatic Identification of Neutral Point Characteristics of Atmospheric Polarization Pattern

Under ideal weather conditions, the angle of polarization image the atmospheric polarization pattern has a significant feature distribution similar to the "∞" glyph. Not only can it effectively reflect the characteristic changes in the atmospheric polarization pattern of the sky light, but also provide a basis for the autonomous orientation method study based on the atmospheric polarization pattern, as shown in Fig. 2.1. One of the important features in the distribution pattern of atmospheric polarization is the neutral point. The neutral point is formed by the polarization singularity in the distribution pattern of sky sunlight due to the depolarization effect from large particles such as aerosols in the atmosphere on the atmospheric polarization pattern

D. Zhao, *Intelligent Information Processing for Polarization Compass and Inertial Integrated Navigation System*, Navigation: Science and Technology 15,
https://doi.org/10.1007/978-981-97-7135-6_2

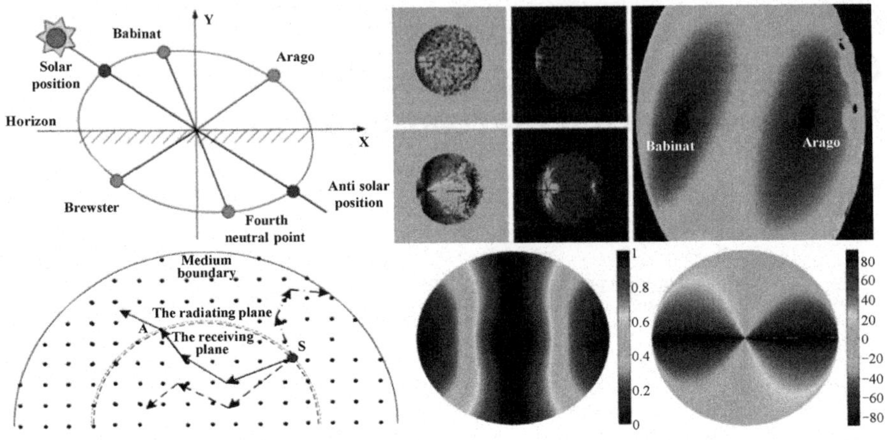

Fig. 2.1 Distribution characteristics of the atmospheric polarization

[1]. The neutral point is also the window of the atmospheric polarization pattern in the sky. In order to obtain accurate and valuable target characteristics, the neutral point can be used for remote sensing target detection by way of providing an accurate directional reference for polarization navigation [2].

In this book, the polarization directions of a pixelated subwavelength metal polarization grating is set as 0°, 45°, 90°, and 135°, respectively. The atmospheric polarization mode characteristics can be expressed using Stokes vectors $S = (I, Q, U, V)$ as:

$$I = \frac{1}{2}\left[I(0^\circ) + I(45^\circ) + I(90^\circ) + I(135^\circ)\right] \tag{2.1}$$

$$Q = I(0^\circ) - I(90^\circ) \tag{2.2}$$

$$U = I(45^\circ) - I(135^\circ) \tag{2.3}$$

where I represents the intensity values of all four pixels in the vicinity of the polarization sensor 2 × 2 pixels. U and Q suggest the linearly polarization components in 0° and 90° directions, as well as in 45° and 135°, respectively. V implies the circular polarization component and it is ignored due to the almost absence in nature. Hence, the degree of polarization (DoP) and angle of polarization (AoP) are expressed mathematically as following.

$$\text{DoP} = \frac{\sqrt{Q^2 + U^2}}{I} \tag{2.4}$$

$$\text{AoP} = \frac{1}{2}\arctan\left(\frac{U}{Q}\right) \tag{2.5}$$

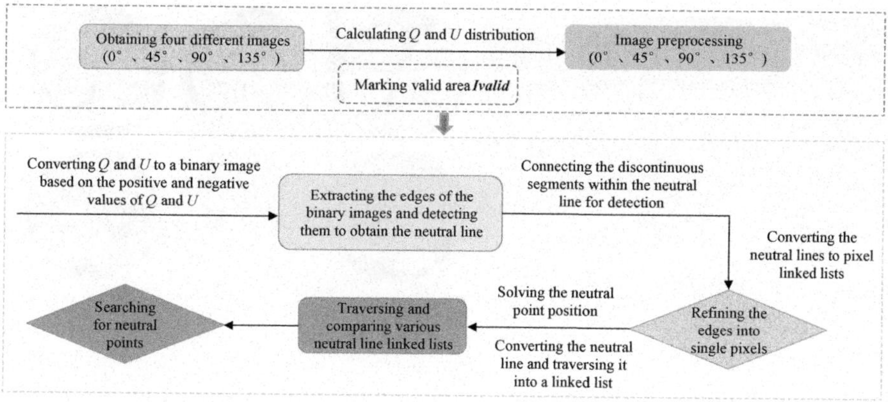

Fig. 2.2 Algorithm flow of the neutral point identification method based on the intersection characteristics of neutral points

DoP in the atmospheric polarization pattern is equal to zero at the neutral point. It can be seen that the linearly polarization components U and Q of the Stokes vector must be equal to zero at the same time according to Eq. (2.4). As a consequence, the neutral line is the set of sky neutral points $U = 0$ and $Q = 0$ in the atmospheric polarization pattern. Since the neutral point needs to satisfy both the characteristics of the linearly polarization component $U = 0$ and $Q = 0$ in the atmospheric polarization pattern, the neutral point is located at the intersection of the neutral line of the linearly polarization component $Q = 0$ and the neutral line of the linearly polarization component $U = 0$. The neutral point of atmospheric polarization can be detected and identified by the intersection characteristics of the neutral line of the linearly polarization component $U = 0$ and the neutral line of the polarization component $Q = 0$, as shown in Fig. 2.2 [3].

At present, the morphological characterization of atmospheric polarization patterns is mainly based on the different particles scattering principle. The characteristic information of atmospheric polarization patterns is relatively scarce. In the practical application of atmospheric polarization pattern, it is necessary to effectively extract the polarization features and parse them into feature patterns that can be recognized by computer hardware [4, 5], which is shown as Fig. 2.3.

2.1.2 *Orientation Algorithm Based on Solar Meridian for Imaging Bioinspired Polarization Compass*

The sky polarization pattern is a special polarization distribution pattern generated by sunlight after it is scattered by atmospheric particles. Consequently, it has a significant distribution law. The scattering particles are mainly composed of atmospheric molecules in clear weather and their size is much smaller than the wavelength of light

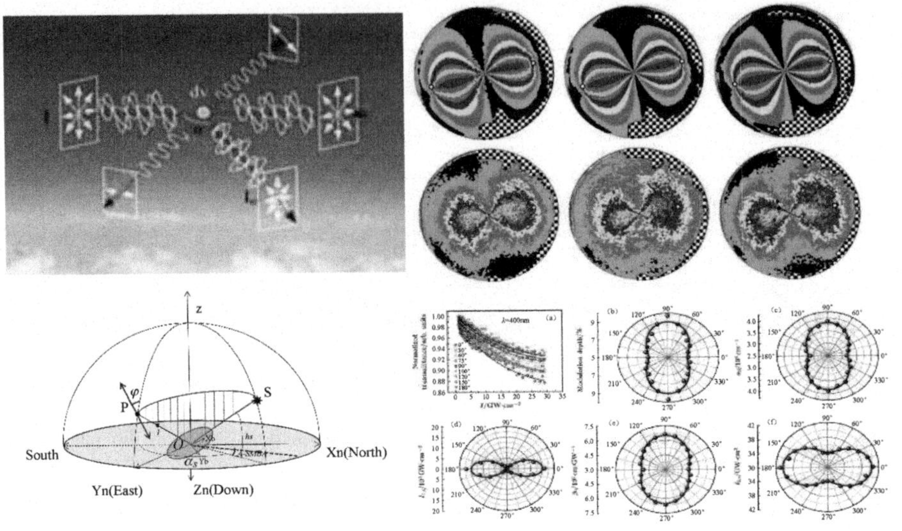

Fig. 2.3 Characterization method of the atmospheric polarization

(about 3 orders of magnitude smaller). The first-order Rayleigh scattering model can be used to describe the atmospheric scattering process in such weather, in which the E-vector direction also known as electrical vibration vector in the light wave of the scattered light is perpendicular to the scattering surface [6].

Figure 2.4 shows a classical 3D horizontal reference coordinate system in relation to the geographical direction. In this system, the coordinate axis direction is taken as east, north and zenith, respectively. The coordinate origin O represents the position of the observer. ϕ is the polarization angle of the incident light and θ is the scattering angle. S represents the sun direction on the celestial sphere, using the zenith angle γ_s and the azimuth angle α_s to describe. The zenith angle and the solar altitude angle are complementary to each other. P implies the observation direction. γ and α are the zenith angle and azimuth angle, respectively.

According to the AoP characteristic information of the atmospheric polarization pattern, the heading angle information of the carrier in the actual motion process is solved in real time. Accordingly, the following coordinate system needs to be defined and converted. In the light of the available references, the right-hand Cartesian coordinate system is defined as follows [7]:

The coordinate system where the polarization camera is located ($OX_cY_cZ_c$): X_c axis and Y_c axis refer to the horizontal and vertical axes of the polarization imager, respectively. Z_c axis is the light axis of the polarization imager and finally points to the zenith direction after the entire polarization imaging system is leveled.

The coordinate system where the incident light is located ($O_iX_iY_iZ_i$): Z_i axis points to the observation direction from the observer. X_i axis is the observer observation direction and is located in the plane (OPP'). X_i, Y_i and Z_i form a right-hand cartesian coordinate system. To simplify the drawing, Y_i axis is not marked in the diagram.

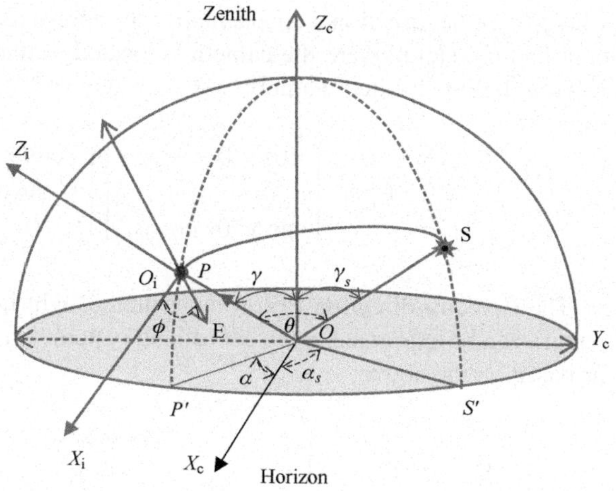

Fig. 2.4 Three-dimensional spatial coordinate system of the atmospheric polarization pattern

According to the first-order Rayleigh scattering model, the DoP in the sky of the atmospheric polarization pattern is expressed as [8]:

$$d = \frac{\sin^2\theta}{1+\cos^2\theta} \tag{2.6}$$

where θ is the scattering angle.

The observer direction vector and the solar direction vector can be expressed in the coordinate system of the polarization camera described above as:

$$\overrightarrow{\mathrm{OS}}^c = \left[\,\sin\gamma_S\cos\alpha_S\ \sin\gamma_S\sin\alpha_S\ \cos\gamma_S\,\right]^{\mathrm{T}} \tag{2.7}$$

$$\overrightarrow{\mathrm{OP}}^c = \left[\,\sin\gamma\cos\alpha\ \sin\gamma\sin\alpha\ \cos\gamma\,\right]^{\mathrm{T}} \tag{2.8}$$

From Eqs. 2.7 and 2.8, the scattering angle θ can be calculated:

$$\cos\theta = \overrightarrow{\mathrm{OS}}^c\cdot\overrightarrow{\mathrm{OP}}^c = \sin\gamma_S\sin\gamma\cos(\alpha-\alpha_S)+\cos\gamma_S\cos\gamma \tag{2.9}$$

The theoretical DoP d can be calculated by Eqs. (2.7) and (2.8). AoP ϕ is defined as the angle between the E-vector direction of the incident light and the axis X_i of the coordinate system of the incident light. According to the first-order Rayleigh scattering model, the E-vector direction $\overrightarrow{\mathrm{PE}}$ of the scattered light in the atmospheric polarization pattern is represented as:

$$\overrightarrow{\mathrm{PE}}^c = \overrightarrow{\mathrm{OS}}^c\times\overrightarrow{\mathrm{OP}}^c = \begin{bmatrix}\sin\alpha_S\sin\gamma_S\cos\gamma-\sin\alpha\cos\gamma_S\sin\gamma\\ \cos\alpha\cos\gamma_S\sin\gamma-\cos\alpha_S\sin\gamma_S\cos\gamma\\ \sin(\alpha-\alpha_S)\sin\gamma_S\sin\gamma\end{bmatrix} \tag{2.10}$$

where $\boldsymbol{C}_c^i$ is the directional cosine matrix. It represents the transition matrix from the coordinate system where the camera is located to the coordinate system where the incident light is located, namely:

$$\boldsymbol{C}_c^i = \begin{bmatrix} \cos\gamma & 0 & -\sin\gamma \\ 0 & 1 & 0 \\ \sin\gamma & 0 & \cos\gamma \end{bmatrix} \begin{bmatrix} \cos\alpha & \sin\alpha & 0 \\ -\sin\alpha & \cos\alpha & 0 \\ 0 & 0 & 1 \end{bmatrix} \tag{2.11}$$

The E-vector direction $\overrightarrow{\mathrm{PE}}$ of the scattered light in the atmospheric polarization pattern can be expressed in the coordinate system, where the incident light is located as based on the above:

$$\overrightarrow{\mathrm{PE}}^i = C_c^i \overrightarrow{\mathrm{PE}}^c = \begin{bmatrix} -\sin(\alpha-\alpha_S)\sin\gamma_S \\ \cos\gamma_S\sin\gamma - \sin\gamma_S\cos\gamma\cos(\alpha-\alpha_S) \\ 0 \end{bmatrix} \tag{2.12}$$

AoP ϕ of the incident light is calculated as:

$$\tan\phi = \frac{\cos\gamma_S\sin\gamma - \sin\gamma_S\cos\gamma\cos(\alpha-\alpha_S)}{-\sin(\alpha-\alpha_S)\sin\gamma_S} \tag{2.13}$$

Therefore, DoP d and AoP ϕ based on the first-order Rayleigh scattering model can be solved according to Eqs. 2.6 and 2.13.

From the above analysis, it can be seen that the atmospheric polarization pattern is a time-dependent stable spatial distribution, which can be solved by acquiring the solar meridian. Using the above-mentioned classical three-dimensional reference coordinate system, the azimuth angle α_b is the angle between the solar meridian and the axis of the carrier body in the coordinate system where the carrier is located. The coordinate system where the north-east earth is located in the reference coordinate system. The azimuth angle α_a is the angle between the solar meridian and the geographic true north, which can be obtained by astronomical almanac. The heading angle $\varphi = \alpha_a - \alpha_b$ is the axis of the carrier body relative to the geographic true north. The carrier heading can be gained by transforming the value range of Eq. (2.14) [9]. The schematic diagram and flow chart of the entire heading angle are obtained as shown in Fig. 2.5a and b.

$$\varphi = \begin{cases} \varphi & \varphi \in \left(0^\circ, 360^\circ\right) \\ \varphi + 2\pi & \varphi \in \left(-360^\circ, 0^\circ\right) \end{cases} \tag{2.14}$$

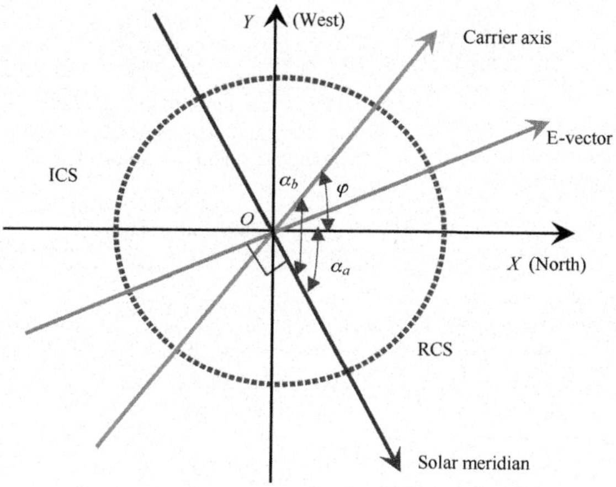

(a) Schematic diagram of calculating heading angle

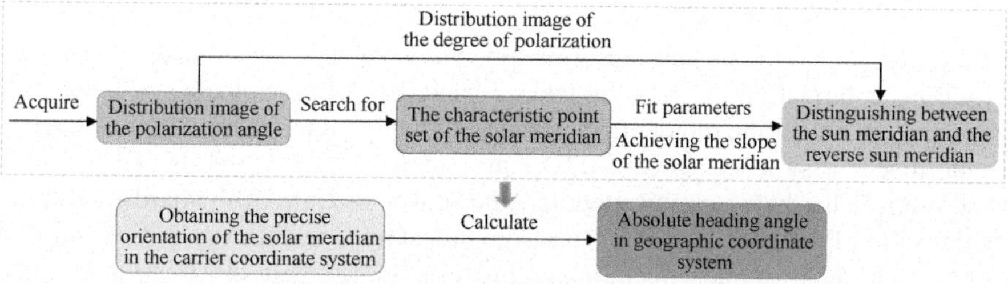

(b) Flow chart of calculating heading angle

Fig. 2.5 **a** Schematic diagram of calculating heading angle, **b** flow chart of calculating heading angle

2.2 Design and Integration for Bioinspired Polarization Compass Based on FPGA

The pixel-level sub-wavelength metal polarization grating with high aspect ratio is integrated with the photosensitive chip and the self-designed FPGA hardware circuit to prepare a bioinspired polarization compass based on the insect polarization orientation mechanism in nature. This compass not only has the advantages of low power, fast speed, compact structure, high resolution, and high sensitivity, but also is able to calculate the carrier heading information in real time. A polarization information acquisition module, a sensor hardware circuit module, and a heading angle data processing module are incorporate in the compass and its functional structure design block diagram is shown in Fig. 2.6.

The functional components such as the pixel-level sub-wavelength metal polarization grating, photosensitive chip, FPGA hardware circuit are reasonably designed

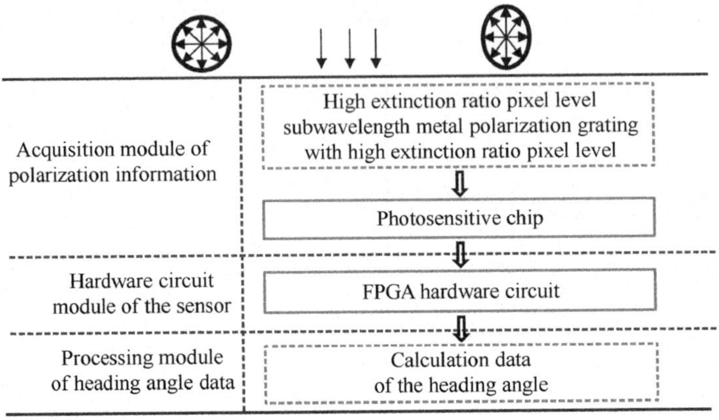

Fig. 2.6 Block diagram of the PC functional structure

and installed into a polarization compass according to the principle of polarization orientation, so that the atmospheric polarization pattern of the sky region can be well achieved and the heading information can be solved. The compass module for acquiring polarization information takes advantage of the photoelectric conversion principle to convert the incident polarization light signal in the sky into an electrical signal that is easy to be processed by subsequent hardware circuits and outputs it to the sensor hardware circuit module. The sensor module of the hardware circuit exploits the FPGA core processor to assign timestamps and data definition information to each data for subsequent signal processing. The data information is output through Ethernet after FPGA decoding, encoding, and analog-to-digital conversion. The test interface provides real-time feedback on the status of the entire system. The processing module for heading angle data mainly makes use of the output signal of the above-mentioned sensing unit through the lower computer and calculates the heading information by the compass.

Specific working principle of the self-designed bioinspired polarization compass based on the FPGA hardware circuit is shown as Fig. 2.7.

Step 1: Obtain the polarization signal in the sky by the high aspect ratio pixel-level sub-wavelength metal polarization grating.
Step 2: Convert the polarization signal into an electrical signal with the sensor chip.
Step 3: Decode the output low-voltage differential electrical signal (LVDS) double-edge data utilizing the programmable logic (PL) terminal in the FPGA circuit.
Step 4: Synchronize the data across the clock domain and convert the data of Step 1 into polarization image data.
Step 5: Encode the image data to the Advanced Extensible Interface (AXI) protocol and write it to the Intellectual Property (IP) core of the Video Direct Memory Access (VDMA).
Step 6: Buffer the polarization image data to DDR3 SDRAM.

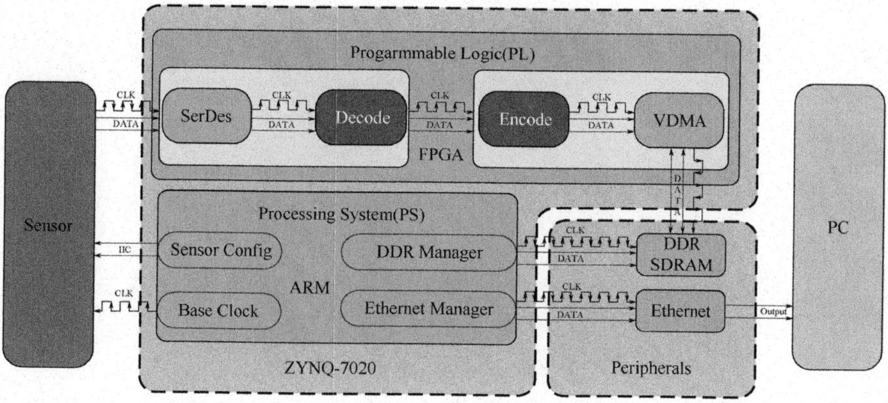

Fig. 2.7 Workflow chart of the bioinspired polarization compass

Step 7: Extract and calculate polarization image data as heading angle data using ZYNQ ARM processing system (PS).

Step 8: Encode the heading data into User Datagram Protocol (UDP) and output it through Ethernet.

Step 9: Decode the output heading angle data and display it on the computer terminal equipment.

The entire bioinspired polarization compass is shown as Fig. 2.8a and the internal FPGA hardware circuit is shown as Fig. 2.8b. The internal polarization imaging system can simultaneously acquire polarization angle images in four directions at one time, which can ensure the synchronization and reliability of image data and can provide a polarization image transmission rate of 30 fps.

2.3 Verification of Bioinspired Polarization Compass Orientation Test

The bioinspired polarization compass self-made in this book is tested by static, turntable and UAV orientation experiments after being completed and integrated. Static and turntable experiment equipment is consisted of a self-made bionic polarization compass, a TBR100 three-axis turntable (Zhuoli Hanguang Company, orientation accuracy is better than 0.05°), a turntable controller, a high-precision fiber optic gyro inertial navigation system/global satellite navigation system (FOG-INS/GNSS) integrated navigation system (IMU-KVH1750/Canada NovAtel PW7720), a computer, a power supply, and so on. The test device is shown in Fig. 2.9 and the turntable is always kept motionless during the static test.

(a) PC Physical diagram

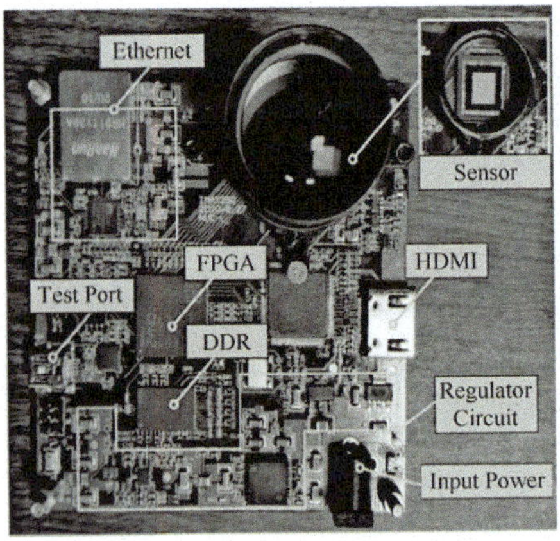

(b) PC Internal hardware circuit diagram

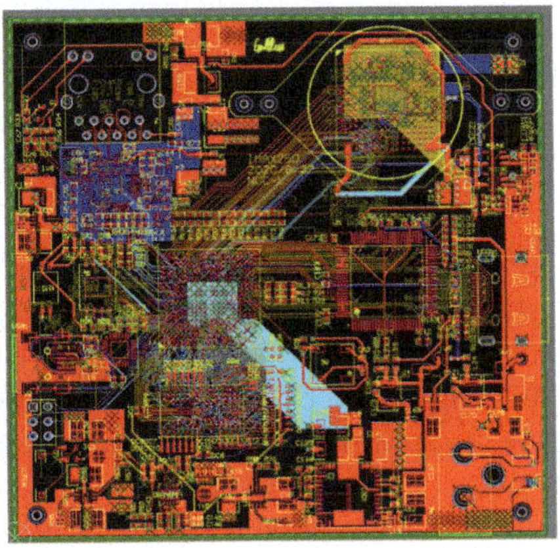

(c) PCB layout

Fig. 2.8 Schematic diagram of the whole bioinspired polarization compass

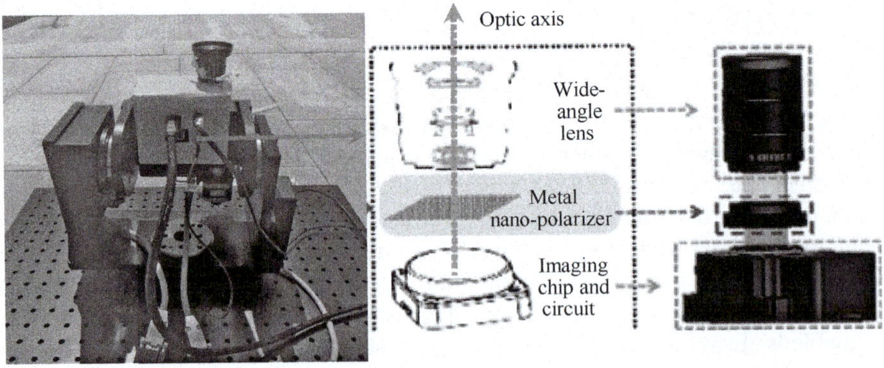

Fig. 2.9 Diagram of the static and turntable test equipment

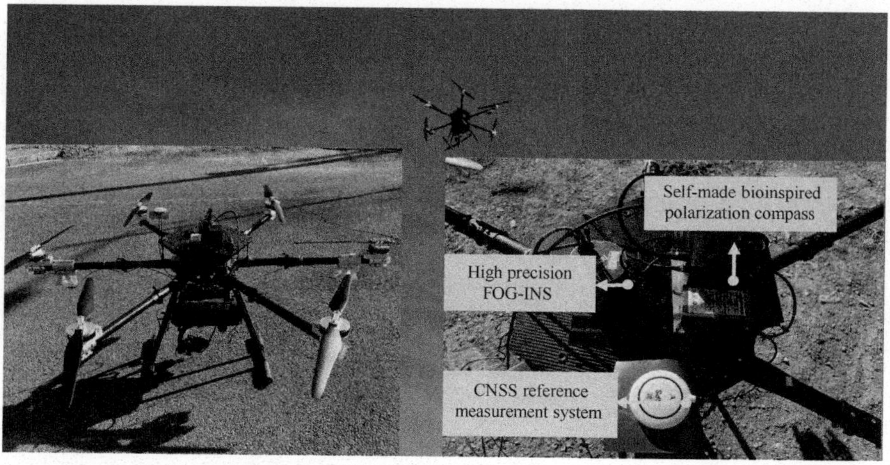

Fig. 2.10 Diagram of the UAV dynamic test device

UAV experiment equipment is comprised of a hexacopter UAV (maximum load: 15kg; model: hexacopter SLM-6S; flight endurance time: 20 min), the same self-made polarization compass and FOG-INS/GNSS measurement reference system as the above-mentioned static test equipment. The whole UAV dynamic test device is shown in Fig. 2.10.

The self-made bioinspired polarization compass and high-precision FOG-INS/GNSS integrated navigation reference are mounted on the TBR100 three-axis turntable and the six-rotor SLM-6S UAV. The installation error is compensated in the solving process considering the influence of installation error on the solution accuracy. The polarization orientation method based on the solar meridian is employed to calculate the heading angle information of the carrier. The orientation accuracy of the bioinspired polarization compass developed in this book is verified by comparing it with the measurement results of the high-precision FOG-INS/GNSS integrated navigation reference measurement. The structure and the whole directional test platform are shown as Fig. 2.11.

Four layers structure is incorporated in the orientation test platform above-mentioned. The first layer is a heading information acquisition device composed of a self-made bioinspired polarization compass and a FOG- INS/GNSS integrated navigation reference. The second layer is the signal processing and control center of the platform system. It can fulfil the function including navigation data acquisition, control carrier turntable movement, communication and data transmission with the ground management computer system by wireless means. The third layer is the static and mobile carrier of the platform system, which conducts static and dynamic tests using a turntable and UAV onboard form. During the static test, the turntable remains stationary. The fourth layer is the ground control system, responsible for sending control instructions to the platform system and storing test data.

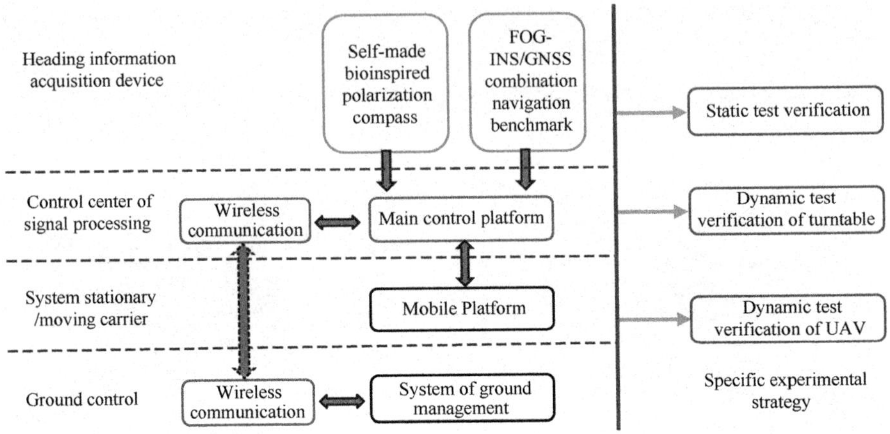

Fig. 2.11 Structure and test scheme of the PC orientation test platform

Four typical indicators, namely mean absolute error (MAE), standard deviation (SD), root mean square error (RMSE), and uncertainty (UC), are introduced to evaluate the orientation performance of the self-made bioinspired polarized light compass.

For the time series $N = (n_1, n_2, \ldots, n_t)$ with a time step t output from a compass, MAE is utilized to estimate the average absolute value of the deviation of all individual observations from the arithmetic mean of the compass output. It avoids the errors from canceling each other out and accurately reflects the magnitude of the actual prediction error. The mathematical formula is expressed as:

$$\text{MAE} = \frac{1}{N}\sum_{t=1}^{N}\left|h_t - \tilde{h}_t\right| \tag{2.15}$$

where h_t and $\tilde{h}_t$ represent the observed value and the simulated value at the time step t, respectively.

SD is applied to measure how discrete each data piece is within a set of datasets. The larger SD suggests the greater the difference between the data in the dataset and the mean of the dataset. Conversely, the smaller SD hints the closer the data in the dataset to the mean of the dataset. The mathematical formula is expressed as:

$$\text{SD} = \sqrt{\frac{1}{N}\sum_{t=1}^{N}\left(h_t - \overline{h}\right)^2} \tag{2.16}$$

where $\overline{h}$ is the average of the observations over the time step t.

RMSE is used to measure the deviation between the predicted value $\tilde{h}_t$ and the true value h_t. The RMSE value is closed to zero, which indicates that the smaller error of the compass and the more accurate the prediction of the heading information. The mathematical formula is expressed as:

$$\mathrm{RMSE} = \sqrt{\frac{1}{N}\sum_{t=1}^{n}\left(h_t - \tilde{h}_t\right)^2} \tag{2.17}$$

UA is an indicator that can be used to measure the quality of measurement results and the degree of reliability of the measured results. It refers to the degree of uncertainty about the measured value due to the existence of measurement error. The smaller UC implies the higher the quality of the measurement results and the higher the usable value of the data. Conversely, the lower UC the usable value of the data means the lower the reliability of the data. There are three types of uncertainty: Class A uncertainty (UA), Class B uncertainty (UB) and synthetic uncertainty (UC). UA is a standard uncertainty evaluated by statistical analysis of observations. UB is a standard uncertainty that is evaluated by statistical analysis different from observations. UC is the standard uncertainty calculated by the variance and covariance of other quantities when the measurement result is obtained from the values of several other quantities. In this book, the Class A uncertainty (UA) is calculated and the mathematical formula is expressed as:

$$\mathrm{UA} = \sqrt{\frac{\sum_{t=1}^{n}\left(h_t - \overline{h}\right)^2}{n(n-1)}} \tag{2.18}$$

2.3.1 *Static Orientation Test*

The static test results are shown as Fig. 2.12a and Table 2.1. The red dotted line stands for the original heading data of the bioinspired polarization compass statically and the solid black line is on behalf of the heading data output by the FOG-INS/GNSS integrated navigation datum. Figure 2.12b shows the heading angle error curve of the bioinspired polarization compass under static test conditions and its static orientation accuracy is 0.1118° (RMSE).

2.3.2 *Turntable Dynamic Orientation Test*

The turntable dynamic test equipment is the same as the above- mentioned static test equipment, where the rotation angle of the turntable is 0°–120° in the whole test process. The test results are shown in Fig. 2.13 and Table 2.2. The red dotted line

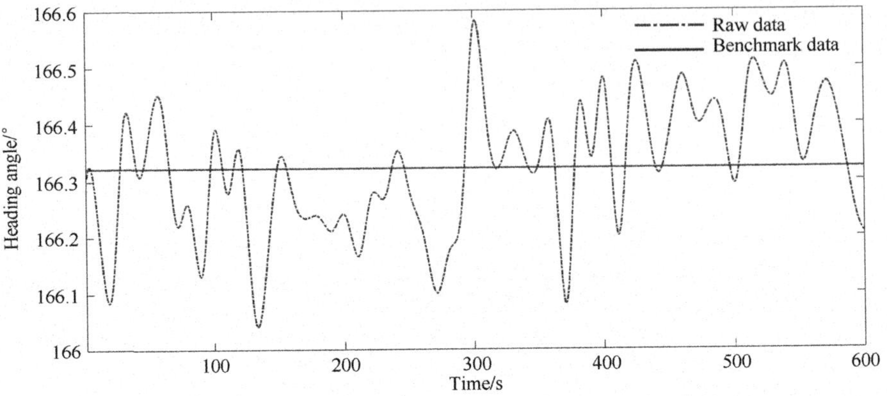

(a) Static test results of the polarizaton compass

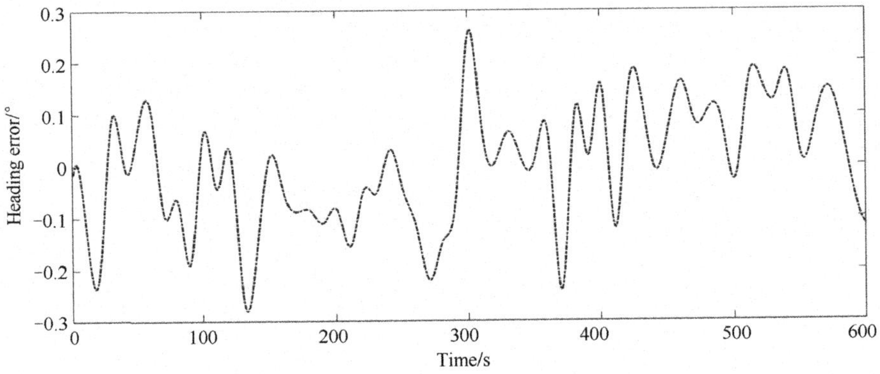

(b) Data diagram of the static heading error of the polatization compass

Fig. 2.12 **a** Static test results of the polarizaton compass, **b** data diagram of the static heading error of the polatization compass

Table 2.1 Heading error indexes obtained from the PC static test

	MFA/°	SD/°	RMSE/°	UA/°
Heading angle error	0.0916	0.1119	0.1118	0.0046

shown in Fig. 2.13a denotes the dynamic output heading angle data of the bioinspired polarization compass. The black solid line is the dynamic output heading angle data of the FOG-INS/GNSS integrated navigation reference turntable. Figure 2.13b shows the heading angle error curve of the bioinspired polarization compass under the turntable dynamic test condition with an orientation accuracy of 0.2081° (RMSE).

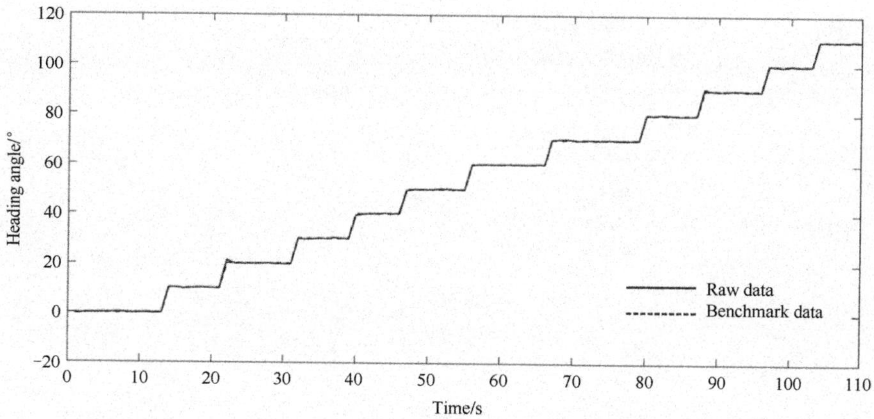

(a) Turntable dynamic test results of the polarizaton compass

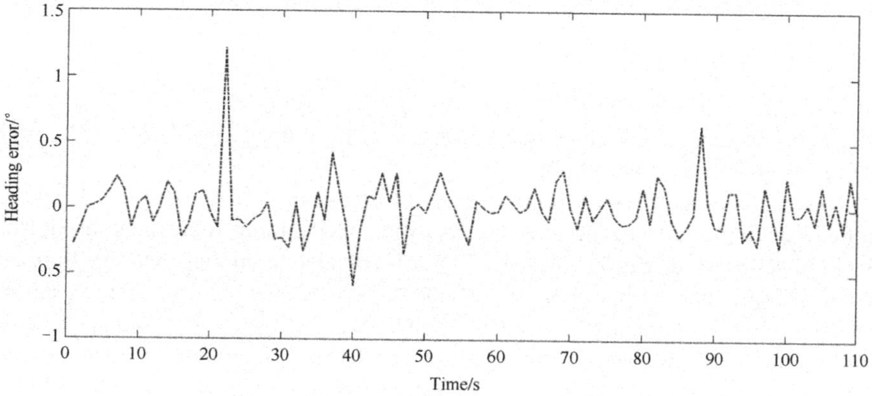

(b) Data diagram of the turntable dynamic heading error for PC

Fig. 2.13 **a** Turntable dynamic test results of the polarizaton compass, **b** data diagram of the turntable dynamic heading error for PC

Table 2.2 Heading error indexes obtained from the PC turntable dynamic test

	MFA/°	SD/°	RMSE/°	UA/°
Heading angle error	0.1443	0.2090	0.2081	0.0200

2.3.3 UAV Dynamic Orientation Test

The UAV airborne test was conducted on the campus of North University of China at sunset (16:10–17:30) on November 12, 2022. During the airborne process, UAV

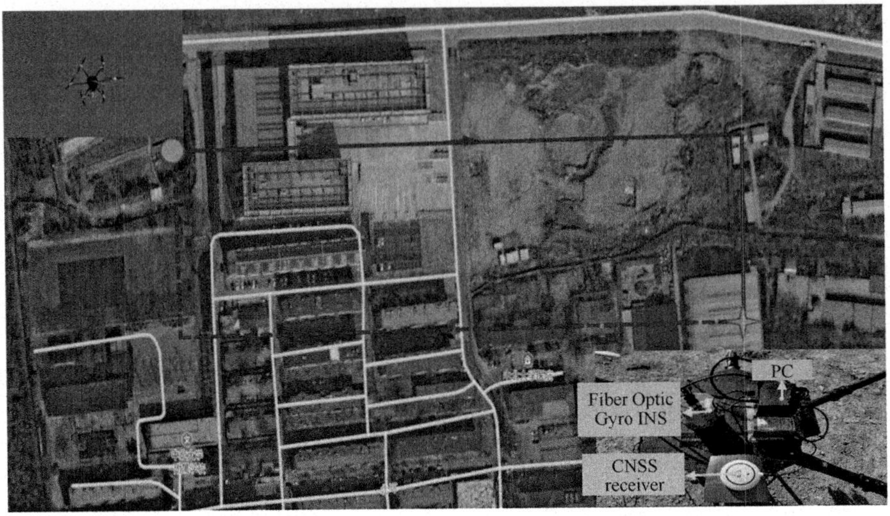

Fig. 2.14 Diagram of the UAV test

flies at an altitude of 310m to the ground and a flight distance of 500m. The flight trajectory is shown as Fig. 2.14.

The results of the UAV airborne test are shown as Fig. 2.15 and Table 2.3. The red dotted line in Fig. 2.15a expresses the UAV dynamic heading angle data output from the bioinspired polarization compass. The solid black line implies the UAV dynamic heading angle data output from the FOG-INS/GNSS integrated navigation datum. The spike in the figures is the result of the obvious change of the compass heading angle when the UAV turns during flight. Figure 2.15b shows the heading angle error curve of the self-made bioinspired polarization compass under the condition of UAV airborne dynamic test and its dynamic orientation accuracy is 1.0900° (RMSE).

In terms of orientation accuracy, the bioinspired polarization compass based on FPGA hardware circuit and high aspect ratio pixel level subwavelength metal polarization grating is compared with existing typical polarization navigation sensor prototypes and microarray polarized light compasses [1, 2]. The results are shown as Table 2.4.

2.4 Chapter Summary

This chapter focuses on the design, integration and test of orientation methods and systems based on the atmospheric polarization pattern. The self-made bioinspired polarization compass is verified by static, turntable dynamic and UAV airborne dynamic orientation tests.

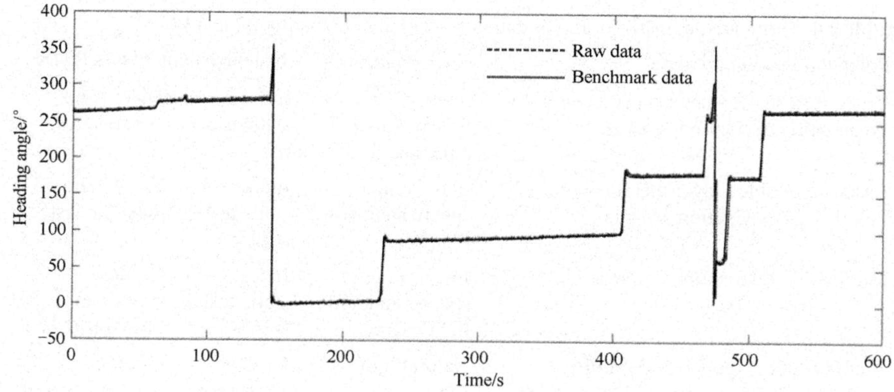

(a) UAV dynamic test results of the polarizaton compass

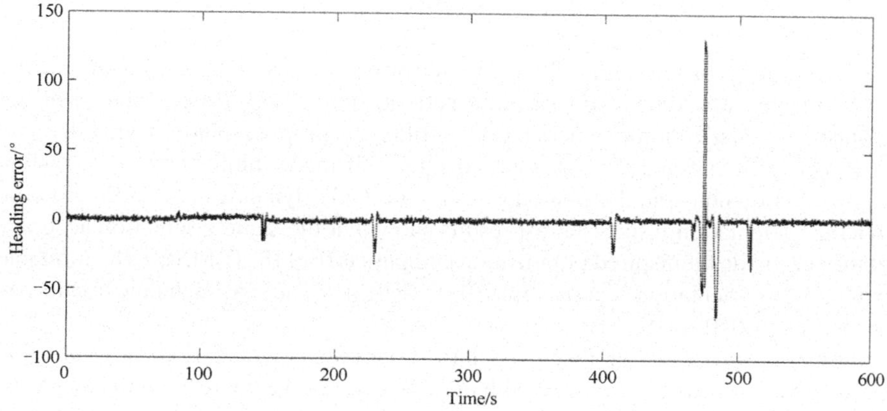

(b) Diagram of the UAV dynamic heading error for PC

Fig. 2.15 **a** UAV dynamic test results of the polarizaton compass, **b** diagram of the UAV dynamic heading error for PC

Table 2.3 Heading error indexes obtained from the UAV dynamic test

	MFA/°	SD/°	RMSE/°	UA/°
Heading angle error	0.8940	1.0905	1.0900	0.1022

Firstly, the neutral point characteristics and automatic identification of atmospheric polarization patterns and the morphological characterization of atmospheric polarization patterns are analyzed, which provides theoretical support for the follow-up research of the solar meridian-based heading angle calculation method.

Table 2.4 Comparison results of the orientation accuracy of existing typical PC

Polarization compass type	Test conditions	Test environment and RMSE	
Based on a single–layer metal nanograting Polarization navigation sensors	Integrating sphere standard light source	Indoor static test 0.8°	There is no dynamic test
Based on double–layer metal nanograting Polarization navigation sensors	Integrating sphere standard light source	Indoor static test 0.7°	There is no dynamic test
Microarray polarization compass	Integrating sphere standard light source	Indoor static test 0.15°	Outdoor single-axis turntable 0.37°
A polarized compass based on FPGA hardware circuitry and a high aspect ratio pixel-level sub-wavelength metal polarization grating	Natural light source	Outdoor static test 0.11°	Outdoor turntable testing 0.21° Outdoor drone airborne 1.09°

Secondly, according to the Stokes vector principle of solving the heading angle based on the atmospheric polarization pattern, the overall design scheme of the bioinspired polarization compass based on FPGA is proposed and integrated.

Finally, the orientation accuracy of the self-made bioinspired polarization compass is demonstrated employing static, turntable dynamics and UAV airborne dynamic orientation tests. The test results show that the static orientation accuracy of the self-made bioinspired polarization compass is 0.1118° (RMSE). The turntable dynamic test orientation accuracy is 0.2081° (RMSE). The UAV dynamic orientation accuracy is 1.0900° (RMSE).

During the process of bioinspired polarization compass orientation, the orientation cumulative error can be corrected with the assistance of the micro-inertial navigation system. The micro-inertial navigation system commonly used in micro-UAVs, such as STIM300 and KY-IMU112, has an optimal orientation accuracy of about 0.5°. The existing bioinspired polarization compass can have an orientation accuracy of better than 0.5° under ideal working conditions on sunny days and can be integrated with micro inertial navigation systems consequently. However, the orientation error of the bioinspired polarization compass will increase rapidly when the carrier is tilted or disturbed by harsh environments such as haze, dust and occlusion, reaching several degrees or even tens of degrees. The orientation error of the micro-inertial navigation system cannot be corrected. Therefore, it is necessary to carry out the research on the bioinspired polarization compass/inertial navigation integrated orientation error processing method for the above-mentioned complex environmental conditions.

References

1. Xindong W, Donghua Z, Hua Y, Tao J, Yongxin L, Chong S. Bionic polarization orientation method under severe weather. Navigat Posit Timing. 2022;9(2):104–11.
2. Rui H. Research on portable measurement method of neutral points in atmospheric polarization pattern. GUIIN UNIVERSITY OF ELECTRONIC TECHNOLOGY; 2020.
3. Rui J, Xia W, Yifan Z, et al. Bionic navigation based on local atmospheric polarization. Acta Aeronautica ET Astronautica Sinica. 2020;41(S2):1–7.
4. Ziqian W, Xudong Z, Haihong J, Zhiguo F. All sky turbid atmospheric polarization pattern modeling based on Monte Carlo method. Chin J Lasers. 2014;41(10):1–9.
5. Jun T, Yubo W, Donghua Z, et al. Application of polarization compass system on solar position calculation. Optik. 2019;187:135–47.
6. Chenguang W, Jun T, et al. Detection and model reconstruction of atmospheric polarization pattern based on Rayleigh scattering theory. J North Univ China. 2015;36(5):570–5.
7. Tao M. Research on position and orientation constraints based bio-inspired navigation. National University of Defense Technology; 2015.
8. Xiong L. Research on modeling and calibration method of a bionic polarization sensor. North China University of Technology; 2019.
9. Shan L, Donghua Z, Hua Y, et al. Three-dimensional attitude determination strategy for fused polarization and geomagnetism. Appl Opt. 2022;61(3):765–74.

Chapter 3
Processing Technology for Bioinspired Polarization Compass Noise

According to the static and dynamic test results for the bioinspired polarization compass, the heading accuracy is seriously affected by the compass noise. This noise comes not only from the AoP image acquired by the compass polarization image sensor, but also from the compass circuit. Therefore, this chapter concentrates the noise analysis and denoising methods for the bioinspired polarization compass. Firstly, different types of noise and their generation mechanism of the bioinspired polarization compass are explored and the noise characteristics are analyzed. Secondly, the denoising algorithm based on Multi-scale Transform (MST) is presented for the bioinspired polarization compass. Finally, the denoising methods for AoP image and heading angle data in significantly improving the orientation accuracy of the polarization compass are verified and analyzed in detail in the way of various experiments.

3.1 Noise Analysis for Bioinspired Polarization Compass

According to the research status of the bioinspired polarization compass orientation error processing method in Sect. 1.2.2 at home and abroad, it can be seen that the existing bioinspired polarization compass orientation methods pay little attention to the research on noise analysis and processing methods, which leads to the serious influence of noise on the heading measurement accuracy. As a result, it is necessary to comprehensively analyze the noise generation mechanism and its characteristics so as to improve the orientation accuracy of the bioinspired polarization compass.

D. Zhao, *Intelligent Information Processing for Polarization Compass and Inertial Integrated Navigation System*, Navigation: Science and Technology 15,
https://doi.org/10.1007/978-981-97-7135-6_3

3.1.1 Analysis of the Generation Mechanism and Characteristics for Polarization Angle Image Noise

According to the polarization imaging principle and heading angle data source, the research on the noise generation mechanism of the bioinspired polarization compass is not only helpful to analyze different noise types and characteristics in AoP images and heading angle data, but also has important guiding significance for selecting and designing appropriate AoP image denoising methods and heading angle data denoising methods.

The polarization imaging system of a bioinspired polarization compass is a monocular image sensor composed of a pixel-level sub-wavelength metal polarization grating with high-aspect ratio, a photosensitive chip, and a FPGA hardware circuit. The imaging principle of the polarization image acquisition system is as follows. The incident light intensity received by the sensor is increased through a lens. Each pixel unit in the pixel-level polarization grating array obtains polarization signals in the 0°, 45°, 90° and 135° directions after being scattered by atmospheric particles. During the exposure process of the image sensor, the photons fall on the reverse bias photodiode so that the reverse voltage on the photodiode decreases. At the end of the exposure, the voltage through the photodiode is measured and read. The photodiode is reset for the next exposure. Eventually, the light signal is converted into an electrical signal and the electrical signal is digitized by an analog-to-digital converter to generate an AoP image.

From the above imaging principles, it can be seen that the noise existed in AoP image is mainly generated by the photon fluctuation, the photoelectric conversion in the pixel circuit, the electron fluctuation, and finally the conversion between the analog signal and the digital signal. The noise generated by photon fluctuations mainly includes shot noise, non-uniform light response noise, dark current non-uniform noise, and crosstalk noise [1]. The specific generation mechanism and characteristic analysis are as follows:

(1) Shot noise

Shot noise n_{shot} is a type of time-dependent random transient noise, including photon shot noise n_{photon} and dark current shot noise n_{dark}. The quantum nature of light itself determines the uncertainty in the number of photons incident on the photodiode. The noise in AoP image due to this fluctuation of the photon count and the conversion of incident photons into electrons by the photodiode and the passage of the photodiode PN junction is called photon shot noise n_{photon}. The magnitude can be expressed as the square root of the charge signal N_{signal} produced in the integration time, i.e.:

$$n_{\text{photon}} = \sqrt{N_{\text{signal}}} \tag{3.1}$$

Dark current shot noise n_{dark} is the noise generated by the thermal motion of photoelectrons. The dark current pixel value of any single frame image in a polarization imaging system will be affected by the relevant shot component. n_{dark} is related to the dark current directly and its magnitude can be expressed as the square root of the number of dark charges N_{dark} produced during the integration time t, i.e.:

$$n_{\text{dark}} = \sqrt{N_{\text{dark}}} = \sqrt{\frac{I_{\text{dark}} \cdot t}{q}} \tag{3.2}$$

where q denotes the amount of charge and I_{dark} describes the value of the dark current. Therefore, n_{shot} can be expressed as:

$$n_{\text{shot}} = n_{\text{photon}} + n_{\text{dark}} \tag{3.3}$$

(2) Noise with non-uniform light response

Owing to the different processing processes of the device, the non-uniform noise of light response is a time-independent fixed-pattern noise generated by the all pixel asynchronous response on the polarization imaging system to the light, which is usually regarded as a pixel gain mismatch phenomenon. This type of noise is becoming more and more significant due to the trend of smaller and smaller pixel sizes in image sensors in recent years, making it increasingly difficult to manufacture imaging systems with the same pixels.

(3) Dark current non-uniform noise

The polarization imaging system needs to be in a long-term exposure mode when collecting polarization images. Hence, different pixels generate different amounts of dark current in the entire pixel array, which is called dark current non-uniform noise. This noise is also a fixed mode noise that is independent of time.

Especially in low light environments, since the dark current noise affects on a small number of pixels in the pixel array, the collected images may show "white spots". Especially when two "white dots" pixels are adjacent in the image, it will produce coupling effects that are difficult to remove. In addition, the dark current noise is closely related to temperature. In crystalline silicon, the dark current noise will double for every 6–8 °C increase in temperature.

(4) Crosstalk noise

Pixel to pixel crosstalk within a polarization imaging system is caused by two different mechanisms: photon crosstalk and propagation crosstalk. Crosstalk causes color mixing and blurring in the image. It also reduces the overall signal-to-noise ratio of the image after color reconstruction in the later stage. Like non-uniform noise in light response, crosstalk noise becomes more and more severe as pixel size decreases.

The noise generated from pixel circuits mainly includes thermal noise, low-frequency flicker noise (also called as $1/f$ noise), reset noise (also called as kTC noise), row noise, and column noise [2]. The specific generation mechanism and characteristics are analyzed as follows.

(1) Thermal noise

Thermal noise is a kind of random transient noise generated by random thermal vibration of electrons in optoelectronic devices. It exists in resistors, source followers, readout transistors, output amplifiers, and analog-to-digital converters in polarization imaging system circuits. The voltage U for thermal noise is expressed as:

$$U = \sqrt{\frac{4kT\alpha}{g}\Delta f} \tag{3.4}$$

where k implies the Boltzmann constant and T is the absolute operating temperature of the polarization imaging system. g represents the MOS transistor transconductance and Δf is the bandwidth. α suggests the correlation coefficient of the MOS transistor operating pattern, which is the constant for a given transistor.

(2) Low-frequency flicker noise

Low-frequency flicker noise, also known as $1/f$ noise, is generated by randomly trapping and diverging carrier regulation channel conductance in silicon crystals. It is a major component of source-following noise. Because the magnitude of low-frequency flicker noise is inversely proportional to the operating frequency of the device, it is also called as $1/f$ noise and its current i is expressed as:

$$i = \sqrt{\frac{K_f}{C_G W_G L_G} \cdot \frac{\Delta f}{f}} \tag{3.5}$$

where K_f indicates the process-dependent constant and f is the device operating frequency. C_G is the gate capacitance per unit area. W_G and L_G represent the gate width and gate length, respectively.

(3) Reset noise

Reset noise is the noise generated by the floating diffuser capacitor node when it is reset. In CMOS image sensors, reset noise appears in the reset phase of the charge detection node [3]. Its charge q_n is expressed as:

$$q_n = \sqrt{kTC} \tag{3.6}$$

where k is the Boltzmann constant and T is the absolute operating temperature of the polarization imaging system. C is the capacitance. As can be seen from Eq. 3.6, the reset noise is only related to the above three parameters, so it is also called as kTC noise.

(4) Line noise

When a given row of pixels is reset in a reset transistor, all pixels in that row are exposed to noise passing through the reset line, transmission gate, or read transistor. This type of noise has a fixed time component and is known as line noise.

(5) Column noise

Column noise is come into being by a mismatch between the sampling and holding capacitors during the pixel circuit reset process. It is also a mismatch between the column amplifier and the perpendicular line of the polarization compass.

(6) Quantify noise

Quantization noise grows out of the analog-to-digital conversion from the polarization imaging system. Digital ramp counters in analog-to-digital converters of pixel circuits typically have a constant step size, resulting in linear quantization noise of the input signal.

3.1.2 Analysis of the Generation Mechanism and Characteristics for Heading Angle Data Noise

The heading angle data is calculated with obtained AoP images from the polarization imaging system in the polarization light compass. It is conveyed through the FPGA hardware circuit. Consequently, the noise arised from the FPGA hardware circuit will have a non-negligible impact on the heading angle data.

FPGA hardware circuit noise mainly includes readout electronic thermal noise, low-frequency flicker noise (also known as $1/f$ noise), random telegram signal noise, and row/column noise. In addition, the analog-to-digital signal conversion in the bioinspired polarization compass circuit also results in quantization noise that cannot be ignored. These types of noise are present in the heading data output from the polarization compass. The row noise that occurs in the polarized compass consists of a time offset and a fixed offset that originates from all given rows of pixels released from the reset. The row noise is generated by the sample/hold capacitor reset and the column amplifier does not match the vertical line of the polarized compass. The noise characteristics of the heading data described above are similar to those yielded by the pixel circuits of the polarization imaging system and will not be repeated here.

The comprehensive noise model for the bioinspired polarization compass is shown as Fig. 3.1 and the specific noise components are summarized in Table 3.1.

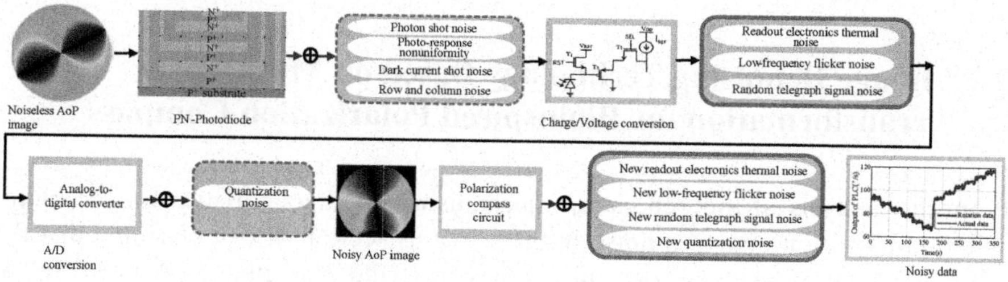

Fig. 3.1 Noise model for the bioinspired polarization compass

Table 3.1 Noise components of the Bioinspired Polarization compass

Serial number	Noise classification	Name of the noise	Sources of noise	Noise impact
1	External noise	External noise	External environmental interference	Polarization angle image
1	Interior noise	Photon shot noise	Uncertainty arises in the number of photons incident on the photodiode	Polarization angle image
2		Dark current shot noise	Photoelectron thermal motion is generated	
3		Noise with uneven light response	The pixel is produced in response to light asynchronously	
4		Dark current non-uniform noise	Produced by long exposures when acquiring images	
5	Interior noise	Crosstalk noise	Interpixel photon crosstalk and propagation crosstalk are generated	
6		Thermal noise	Random thermal vibration of electrons in optoelectronic devices is generated	
7		Low-frequency flicker noise	Random capture and divergence carriers regulate channel conductance generation	
8		Reset noise	Generated when the floating diffusion capacitance is reset	
9		Line noise	Generated when a pixel line is exposed through reset lines, transmission gates, etc.	Heading data
10		Column noise	A mismatch between the sampling and holding capacitances occurs during the pixel circuit reset	
11		Quantize noise	Polarization compass is generated by modulus-to-digital conversion	

3.2 Image Denoising Technology Based on Multi-Scale Transformation for Bioinspired Polarization Compass

From the analysis of the generation mechanism and characteristics for the bioinspired polarization compass noise in the previous section, it can be found that the polarization compass noise not only comes from the AoP image obtained by the polarization imaging system, but also the heading angle data is seriously affected by the compass circuit noise, both of which ultimately leads to a significant reduction in the orientation accuracy of the bioinspired polarization compass.

With regard to image noise, image denoising methods derived from mathematical and statistical development have been widely used, such as total variation method (TV), non-local filtering method (NLM), sparse filtering method (SFM), empirical pattern decomposition method (EMD), and so on. Rudin et al. [4] proposes a total variational (TV) denoising method. This method assumes that the image consists of a small set of piecewise constants, a small number of discontinuous points and edges, and then uses a constraint optimization model to remove image noise. The denoising method has a good denoising effect on segmented constant images, but it has a step effect on imprecise constant images. To this end, Abergel et al. [5] proposed total variational (TV) methods in integrated with other algorithms for denoising. Nevertheless, these methods are only effective for special cases of additive Gaussian noise and Poisson noise. In order to expand the range of image noise removal types, Buades et al. [6] presents a denoising algorithm based on non-local filtering (NLM) and the denoising effect on images with special targets is susceptible to the aperture effect of noise. Deledalle et al. [7] improves NLM denoising methods to smooth the image but leads to slow to run. Compared with NLM, the sparse filtering method (SFM) first transforms the noisy image and denoises the transformed image subsequently. The two most representative methods in SFM are the wavelet soft threshold method and the three-dimensional block matching filtering method (BM3D). The image quality after wavelet soft threshold denoising is generally good, but large artifacts can occur and especially near the edges of the denoised image [8]. The BM3D image denoising method is denoised by enhancing the sparse representation in the transformation domain [9], which is mainly effective in suppressing additive Gaussian white noise. Different from the traditional image denoising methods mentioned above, Huang et al. [10] exploits an empirical pattern decomposition (EMD) method, which adaptively decomposes the signal into eigenpattern functions (IMFs) according to its different frequencies in the physical domain and estimates the IMFs through a screening iterative process. EMD provides a powerful tool for adaptive multiscale analysis of non-stationary signals and it is widely used in various fields.

With regard to the denoising of polarization imaging systems, the latest progress is the denoising of grayscale and monochrome images through color filtering images based on principal component analysis (PCA). Another is the segmentation denoising method for AoP images by subfocal plane polarimeter (DoFP) [11]. The signal-to-noise ratio, theoretical estimation, and measurement results based on Stokes parameters in a subfocal plane polarimeter (DoFP) are analyzed in Ref. [12]. In Ref. [13], the Poisson shot noise of the Stokes polarimeter is minimized by using the noise variance equalization feature independent of the input polarization state. Gaussian noise processing methods in DoFP polarized images are mainly the learning method based on the residual dense network [14], such as the block matching and BM3D filtering algorithm [15], the K-order singular value decomposition (K-SVD) [16], and the integrated of BM3D and KSVD [17], and so on.

From the above analysis, it can be seen that although there are various image denoising methods, the existing image denoising methods are all aimed at suppressing a certain type of noise and lack the comprehensive processing of various types of noise. There is rare research in the heading angle data denoising method for the bioinspired polarization compass. Therefore, based on the noise analysis for the bioinspired polarization compass in previous section, this book proposes a new denoising method for AoP image and heading angle data based on Multi-scale Transform (MST).

3.2.1 Denoising Technology for Polarization Angle Image Based on Multi-Scale Transformation

A new denoising method for AoP images based on Multiscale Transformation (MST) firstly adopts the two-dimensional Empirical Mode Decomposition (BEMD) method to adaptively decompose the original noisy polarization angle image into multiple different two-dimensional image intrinsic mode functions (Bi-dimensional) through continuous iterative screening intrinsic mode functions (BIMF). The corresponding image denoising algorithm is exploited to suppress the noise in the decomposed AoP images with BEMD.

Two-dimensional empirical pattern decomposition (BEMD) proposed by J.C. Nunes et al. in 2003 is a two-dimensional form of empirical pattern decomposition (EMD). EMD does not need to preset or give a basis function, which is an adaptive multi-scale decomposition method for image signals. It decomposes nonlinear and nonstationary signals into multiple frequency IMFs according to the characteristics of a given input signal itself [18]. Therefore, BEMD, as an EMD extension technology, is also an image analysis technology that can be driven by the input signal to adaptively decompose a given image signal into a series of BIMFs components and residual components. Various BIMF components represent the specific details of the image signal at different scales and the residual component denotes the contour information of the image signal. Low-order BIMFs suggest high-frequency noise pattern and high-order BIMFs represent low-frequency signal pattern. The local orthogonal between each component BIMFs can effectively reflect the original image feature information. Therefore, the algorithm has been widely used in remote sensing, signal analysis, biomedicine, and integrated navigation.

For a given $m \times n$ two-dimensional noise image $f(x, y)$, the BEMD decomposition principle is expressed as follows [19]:

Step 1: Initialize $m \times n$ the image signal $f(x, y)$.

$$F_{i,j-1}(x, y) = f(x, y), i = 1, j = 1, (x, y) \in [0, m] \times [0, n] \tag{3.7}$$

Step 2: Calculate $F_{i,j-1}(x, y)$ maximum point sets $\max_{i,j-1}(x, y)$ and minimum point sets $\min_{i,j-1}(x, y)$.

Step 3: Obtain $F_{i,j-1}(x, y)$ the upper envelope surface $eu_{i,j-1}(x, y)$ and the lower envelope surface $el_{i,j-1}(x, y)$ according to the interpolation fitting of the maximum point set $\max_{i,j-1}(x, y)$ and the minimum point set $\min_{i,j-1}(x, y)$.

Step 4: Calculate the average envelope value $E_{i,j}(x, y)$ based on the upper and lower envelope surfaces.

$$E_{i,j}(x, y) = \frac{eu_{i,j-1}(x, y) + el_{i,j-1}(x, y)}{2} \tag{3.8}$$

Step 5: Calculate the new intermediate variable $N_{i,j-1}(x, y)$.

$$N_{i,j-1}(x, y) = F_{i,j-1}(x, y) - E_{i,j-1}(x, y) \tag{3.9}$$

Step 6: Determine $N_{i,j-1}(x, y)$ whether meets the termination criterion SD for screened BIMF.

$$\mathrm{SD} = \frac{\sum_{x=1}^{m}\sum_{y=1}^{n}\left|N_{i,j-1}(x, y) - F_{i,j-1}(x, y)\right|}{\sum_{x=1}^{m}\sum_{y=1}^{n}\left|F_{i,j-1}\right|^2} \tag{3.10}$$

If SD $< \varepsilon$ (ε usually takes a value of 0.2–0.3), $N_{i,j-1}(x, y)$ acquires the ith BIMF, and that is $\mathrm{BIMF_i}(x, y) = N_{i,j-1}(x, y)$; otherwise update $j = j+1$ and $F_{i,j-1}(x, y) = N_{i,j-2}(x, y)$ by means of repeating Step 1–Step 6.

Step 7: Calculate the new residual component $r_{i,0}(x, y)$.

$$r_{i,0}(x, y) = r_{i-1,0}(x, y) - N_{i-1,j-1}(x, y) \tag{3.11}$$

When the number of extreme points of the new residual component $r_{i,0}(x, y)$ is less than three or the constraints required by each BIMF are reached, the final residual component $r(x, y) = r_{i,0}(x, y)$ is obtained and the BEMD decomposition process is completed; otherwise Step 1–Step 7 is repeated. Each BIMF needs to meet two constraints: (1) the total number of BIMF extrema should be equal to the number of zero crossings, or a difference of up to 1; (2) the mean of the upper and lower envelopes defined by the local extrema at any point of the BIMF should be equal to zero. The result of the decomposition of the entire given $m \times n$ 2D noise image $f(x, y)$ is expressed as:

$$f(x, y) = \sum_{i=1}^{L} \mathrm{BIMF}_i(x, y) + r(x, y) \tag{3.12}$$

where $\mathrm{BIMF}_i(x, y)$ is the ith BIMF component obtained by BEMD decomposition, which is the L total number of BIMF components.

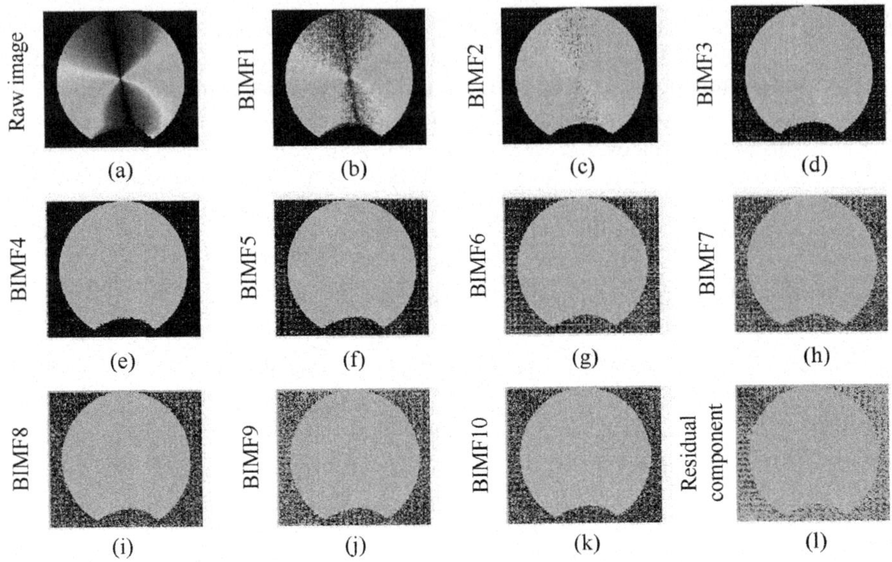

Fig. 3.2 BEMD decomposed view of AoP

Taking the AoP image studied in this book as an example, the BEMD decomposition diagram is shown as Fig. 3.2. Figure 3.2a suggests the source image and Fig. 3.2b–k imply the high-frequency to low-frequency components of BIMF. Figure 3.2l shows the residual component.

3.2.2 MS-PCA Image Denoising Technology Based on BEMD for Bioinspired Polarization Compass

From the analysis in the previous section, it can be seen that the new AoP image denoising method based on multi-scale transform (MST) can adaptively decompose the original noise image into multiple BIMFs with high to low frequency distribution and a residual component through BEMD. In this section, a multi-scale principal component analysis (MS-PCA) denoising method based on BEMD decomposition is proposed for the original noisy AoP image.

Principal component analysis (PCA) is a data dimensionality reduction algorithm, which calculates the eigenvalues and eigenvectors of the covariance matrix for the n-dimensional data matrix. PCA approach selects the eigenvectors corresponding to k large eigenvalues to form a k-dimensional matrix, which is called the principal component. Namely, the new k-dimensional feature data is constructed on the basis of the original n-dimensional feature data and finally reduces the data dimensionality [20]. This method is widely used in image data denoising. Inspired by this, PCA denoising method for AoP images based on BEMD is for further research in this book.

The core steps of the method include: BEMD decomposition, BIMFs classification, threshold selection of PCA image denoising algorithm and so on. The whole AoP image denoising process are as follows:

Step 1: Decompose the original AoP noise image utilizing BEMD.

The BEMD-based MS-PCA polarization angle image denoising method first uses the BEMD image decomposition method introduced in Sect. 3.2.1 to decompose the original AoP noise image $f(x, y)$ into multiple BIMFs with different frequencies and a residual component r(x,y), namely:

$$f(x, y) = \sum_{i=1}^{L} \text{BIMF}_i(x, y) + r(x, y) \tag{3.13}$$

Step 2: Classify the BEMD decomposed results.

According to the content of this study, it is necessary to extract valuable solar meridian information and reduce the amount of denoising calculation for each BIMF. Accordingly, one-dimensional image entropy (1D-IE) is presented as a classification criterion to divide BIMFs into solar meridian-dominant BIMFs (S-BIMFs) and non-solar meridian-dominated BIMFs (NS-BIMFs). This classification method is based on the similarity of the 1D-IE intervals because the 1D-IE values change sharply when the solar meridian information disappears from the BIMF. The 1D-IE values with continuous similarity characteristics from high frequency to low frequency are regarded as solar meridian dominant BIMFs (S-BIMFs). The rest of the BIMFs are regarded as non-solar meridian dominant BIMFs (NS-BIMFs).

Image entropy is a statistical form of image features that reflects the average amount of information in an image. One-dimensional image entropy (1D-IE) represents the amount of information involved in the aggregation features of the gray distribution in the image. Its mathematical formula is expressed as follows:

$$H = \sum_{i=0}^{255} P_i \log_2 P_i \tag{3.14}$$

where P_i represents the proportion of pixels in the image that have grayscale values. Higher entropy H means more information about the image outline. Conversely, H lower means less useful signal.

Step 3: Select the threshold for the AoP image denoising algorithm adaptively.

In order to effectively remove the noise in the AoP image and maximize the information of the solar meridian dominant BIMFs (S-BIMFs), the PCA image denoising algorithm based on principal component contribution rate (R_P) is applied to denoise the S-BIMFs at various scales after 1D-IE classification. The principal component contribution rate (R_P) is determined by the quotient ratio of the eigenvalues (λ_i) of

Table 3.2 Flow chart of the MS-PCA algorithm

Algorithm 1: MS-PCA
Requirement: Polarized image signal $f(x, y)$ **Decompose** 1: BEMD decomposes the original polarization angle image $f(x, y)$ into L BIMFs and a residual component $r(x, y)$ Classify
2: for $i = 1$ to L do
3: Calculate the one-dimensional image entropy (ID-IE) of each BIMF according to Eq. 3.14
4: Divide BIMFs into two categories: S-BIMFs and NS-BIMFs **Denoising** 5: Calculate the principal component contribution rate R_P 6: R_P PCA image denoising algorithm based on the determination of S-BIMFs at each scale 7: Different PCAs remove different S-BIMFs noise and discard NS-BIMFs 8: Reconstruct the denoised S-BIMF 9: End

each BIMF image covariance matrix to the sum of the eigenvalues ($\sum_{i=1}^{L} \lambda_i$) of the entire BIMF image covariance matrix, i.e., $R_P = \lambda_i / \sum_{i=1}^{L} \lambda_i$ [21].

Step 4: Denoise the AoP image.

According to the principal component contribution rate (R_P) calculated in Step 3, the PCA denoising algorithm is used to adaptively remove the intrinsic noise of S-BIMFs at each scale and NS-BIMFs are discarded.

Step 5: Reconstruct the denoised BIMFs image.

The denoising S-BIMFs are reconstructed to obtain the final denoised AoP image. The entire MS-PCA polarization angle image denoising algorithm flow is shown as Table 3.2 and Fig. 3.3.

3.2.3 *Verification of MS-PCA Polarization Angle Image Denoising Method Based on BEMD*

In order to comprehensively analyze the performance of the MS-PCA polarization angle image denoising method proposed in this book, the algorithm performance is evaluated by using two typical indexes (i.e., standard deviation (SD), mean gradient (MG)) for the polarization angle image before and after denoising. The efficiency of the algorithm is evaluated by using two typical indexes, namely, the time complexity (TC) and the spatial complexity (SC). In this study, the performance of the proposed denoising method for AoP images cannot be evaluated by the index of signal-to-noise ratio (SNR) without adding additional noise.

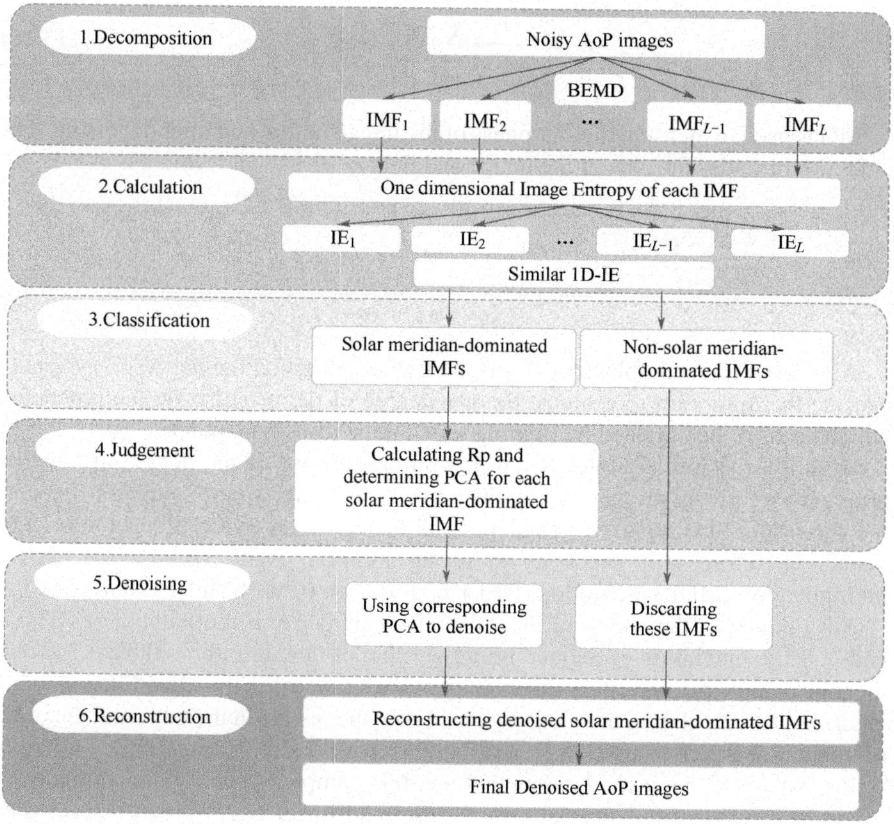

Fig. 3.3 Flow chart of the MS-PCA algorithm

For a given $m \times n$ 2D noisy image $f(x, y)$, the standard deviation (SD) is the dispersion degree between the gray value of the AoP image and the gray value of the average AoP image. The smaller SD indicated the smaller difference between the gray value of the polarization angle and the gray value of the average polarization angle image. The mathematical formula is expressed:

$$\text{SD} = \sqrt{\frac{1}{mxn}\sum_{x=1}^{m}\sum_{y=1}^{n}[G(x, y) - E]} \tag{3.15}$$

where $G(x, y)$ represents the grayscale value of the polarization angle image coordinate points (x, y). E hints the grayscale value of the average polarization angle image, which is expressed as:

$$E = \frac{1}{m \times n} \sum_{i=1}^{m} \sum_{j=1}^{n} G(x, y) \tag{3.16}$$

MG is used to measure the sharpness of the image before and after denoising. The larger MG implies the clearer the image and the mathematical formula is expressed:

$$\mathrm{MG} = \sqrt{\frac{\sum_{x=1}^{m-1} \sum_{y=1}^{n-1} \left[(G(x, y) - G(x+1, y))^2 + (G(x, y) - G(x, y+1))^2 \right]}{2 \times (m-1) \times (n-1)}} \tag{3.17}$$

Algorithm time complexity (TC) is the time required by the hardware system to execute the algorithm to evaluate the use degree of the program on the processor, which needs to be obtained by running tests on the lower computer.

Algorithm spatial complexity (SC) is an assessment of the degree to which a program uses processor memory by the amount of storage space required to execute the algorithm. In complexity calculation, all the operators involved need to be estimated. Therefore, it is necessary to set up addition (ADD), subtraction (SUB), multiplication (MUL), definition (DEF), comparison (CMP), and division (DIV).

S is defined as the length of the AoP image signal, K is the number of BIMFs, and N is the maximum cyclic and iterative value of the algorithm. Table 3.3 shows the time complexity and space complexity of each process in the MS-PCA denoising algorithm by calculating them one by one. From the results in the table and the order of magnitude relationship $O(N!) > O(N^3) > O(N^2) > O(2N \log_2 2N) > O(N) > O(\log_2 2N) > O(1)$, it can be seen that the time complexity and spatial complexity of the MS-PCA are polynomial $O(N^2)$-order and linear $O(N)$-order, respectively. Therefore, the MS-PCA image denoising method is an effective polynomial time-solvable algorithm.

In order to verify the effectiveness and practicability of the proposed MS-PCA polarization image denoising method, experimental verification is carried out in sunny, bad sand and dust weather conditions, respectively.

(1) Sunny weather test verification

Table 3.3 Time and space complexity of the image denoising algorithm based on MS-PCA

Function	T	M (float)
Initialize	$11\mathrm{DEF} \cdot S$	$3 + 4 \otimes S + 4 \otimes 2S$
Local extreme points	$O(N^2)$	$15S + 11 + 2N \otimes S$
Envelope estimation	$3\mathrm{DEF} + (3\mathrm{CMP} + 4\mathrm{ADD} + 2\mathrm{SUB} + 5\mathrm{DEF}) \cdot K \cdot N \cdot S$	$5 + 4N \otimes S$
Envelope mean $M_{i=1}$	$(6\mathrm{CMP} + 5\mathrm{ADD} + 2\mathrm{SUB} + 1\mathrm{DIV}) \cdot K \cdot N \cdot S$	$28 + 7 \otimes S$
Iterative screening	$(2\mathrm{ADD} + 6\mathrm{DEF} + 4\mathrm{CMP}) \cdot N \cdot S$	$4 + N \otimes K$
Complexity	$O(N^2)$	$O(N)$

The test was conducted at the campus football field on May 7, 2023 (15:30–17:00). The test system is composed of a self-made bioinspired polarization compass, TBR100 three-axis turntable, and high-precision light fiber/satellite (FOG-INS/GNSS) integrated navigation system (IMU-KVH1750/Canada NovAtel PW7720) as the measurement reference (reference course accuracy 0.035°). The whole test device and its characteristics are shown as Fig. 3.4 and Table 3.4.

The original AOP image acquired by the polarization imaging system in the above test system is adaptively decomposed into 10 BIMFs (i.e., BIMF1, BIMF2, BIMF3, …, BIMF10) and a residual component by BEMD. The decomposed results of the whole polarization angle image are shown as Fig. 3.5. Obviously, the solar meridian contained in BIMF1 is much clearer than that of BIMF2 and the information from BIMF3 to the residual solar meridian is gradually reduced.

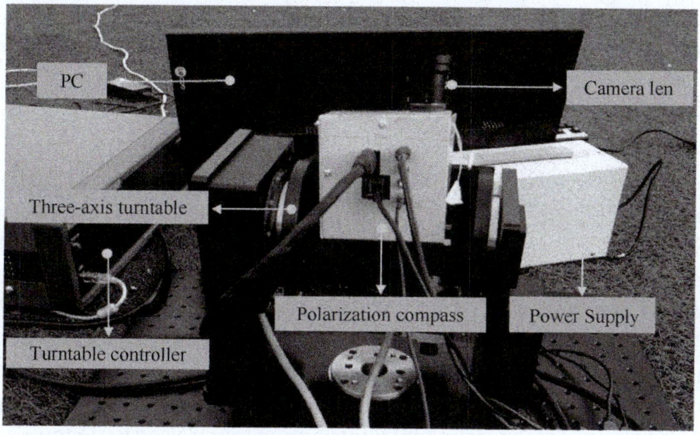

Fig. 3.4 Diagram of the static test equipment in sunny days

Table 3.4 Parameters of the test equipment

Bioinspired polarization compass	Polarization measurement unit	High aspect ratio pixel-level Subwavelength metallic polarization gratings
	Photoreceptor chip	IMX250LQR
	Hardware circuits	FPGA hardware circuitry
FOG-INS/GNSS Measurement datum	Model	KVH 1750
	Orientation accuracy	0.035
TBR100 three-axis turntable	Model	Zhuoli Hanguang TBR100
	Full Step Resolution/°	0.01
	Maximum speed/(°/s)	20
	Orientation accuracy/°	≤0.05

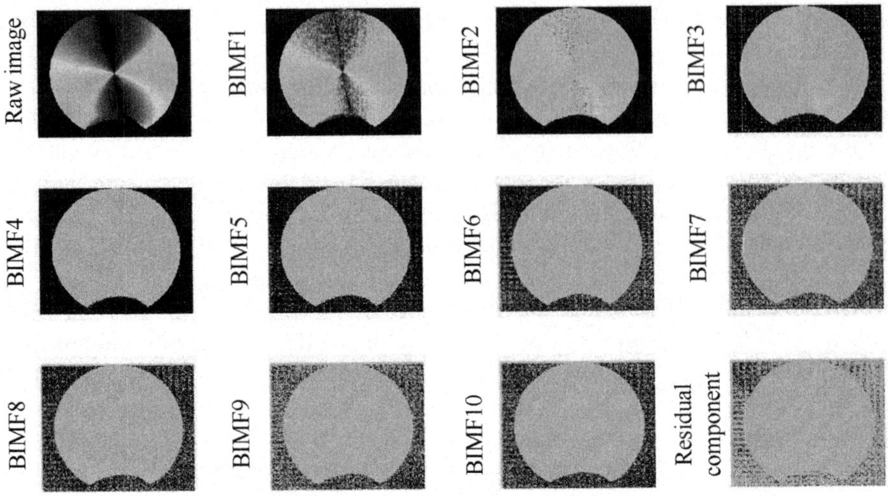

Fig. 3.5 Original and BEMD image of Aop on sunny days

In order to effectively employ one-dimensional image entropy (1D-IE) to divide the decomposed BIMFs and residual components into solar meridian dominant BIMFs (S-BIMFs) and non-solar meridian dominant BIMFs (NS-BIMFs), the 1D-IE value of each BIMF and residual component after decomposition is first calculated and the results are shown in Table 3.5. Based on the similarity law of 1D-IE value intervals, BIMF1 to BIMF5 in the BIMFs sequence are classified as solar meridian dominant BIMFs (S-BIMFs). BIMF6 to the residual components are classified as non-solar meridian dominant BIMFs (NS-BIMFs). Finally, the PCA image denoising method based on principal component contribution rate is used to denoise each BIMF scale of S-BIMFs and the NS-BIMFs are removed. The complete parameter settings of the MS-PCA polarization image denoising algorithm are shown in Table 3.6. The denoising results of the original AoP image are shown in Fig. 3.6.

Table 3.7 shows the comparison results of MG and SD for the MS-PCA polarization angle image denoising method proposed in this book with other existing image denoising techniques (such as BM3D, PDRDN and BM3D-KSVD) for the four original polarization angle images acquired at 0°, 45°, 90° and 135°. As can be seen from the table, the MS-PCA image denoising algorithm proposed in this study achieves the best values in both SD and MG evaluation indexes, indicating that the proposed MS-PCA has the best performance in image denoising.

Figure 3.7 shows the results of a comparative test of the polarization angle error calculated after removing the AoP image noise from the original image acquired by static test tests and according to the proposed MS-PCA and other image denoising methods. It can be seen from the heading angle error curves in the figure that the MS-PCA image denoising method proposed in this book decomposes the original image by BEMD and then performs adaptive PCA denoising for NS-BIMFs, which finally effectively reduces the heading error.

Table 3.5 1D-IE value of the each BIMF and residual component after BEMD on sunny days

	BIMF1	BIMF2	BIMF3	BIMF4	BIMF5	BIMF6	BIMF7	BIMF8	BIMF9	BIMF10	Residual component
1D-IE	0.9056	0.3199	0.3020	0.0284	0.0778	0.4102	0.3125	0.3785	0.3455	0.3316	0.3039

Table 3.6 Parameters of the image denoising algorithm based on MS-PCA on sunny days

MS-PCA parameters	BIMF1		BIMF2		BIMF3		BIMF4		BIMF5	
	λ_1	R_{p1}	λ_2	R_{p2}	λ_3	R_{p3}	λ_4	R_{p4}	λ_5	R_{p5}
Parameter value	0.0929	0.1107	0.0918	0.1093	0.0988	0.1177	0.0942	0.1123	0.0905	0.1078

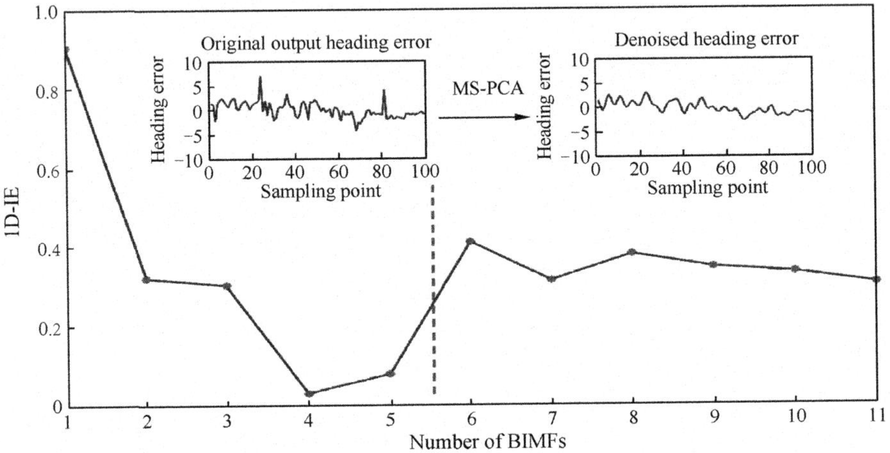

Fig. 3.6 Classification and the results of BIMFs after denoising by MS-PCA

Table 3.7 Comparison after various image denoising algorithms on sunny days

Image	Index	Original image	BM3D	PDRDN	BM3D-KSVD	MS-PCA
1	SD	24.2986	22.8239	21.1726	21.1337	21.0886
	MG	0.1429	0.5720	0.3466	1.1201	1.1254
2	SD	24.3252	22.8836	21.2087	21.1779	21.1009
	MG	0.1426	0.5768	0.3607	1.0249	1.1263
3	SD	24.2494	20.6396	21.1557	21.1198	21.0363
	MG	0.1615	0.6755	0.3507	1.0503	1.1437
4	SD	24.2953	22.8207	21.2013	21.1619	21.0864
	MG	0.1422	0.5736	0.3403	1.0109	1.1201

Table 3.8 compares the mean (Mean), standard deviation (SD) and root mean square error (RMSE) of the static heading error data obtained by various image denoising algorithms. As can be seen from the table, BM3D can remove part of the AoP image noise but the SD and RMSE of the denoising results are still high, indicating that BM3D has limited ability to suppress the AoP image noise. Although the denoising performance of PDRDN and BM3D-KSVD is better than that of BM3D, the heading error is calculated Mean, SV and RMSE are 0.0730°,

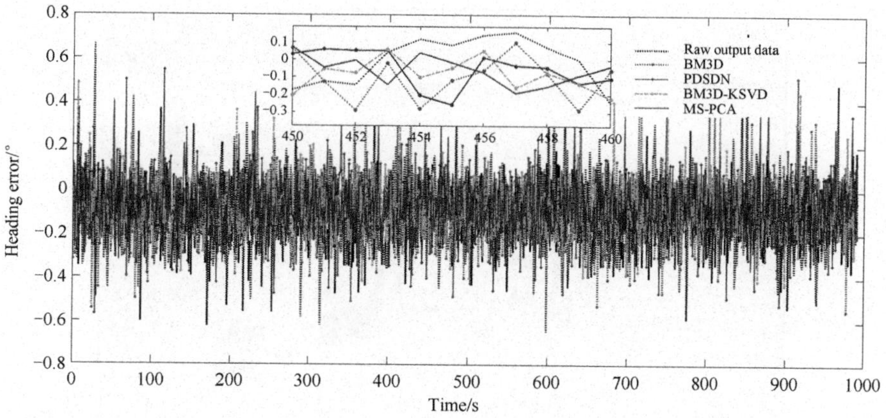

Fig. 3.7 Heading error after various AoP denoising algorithms in static test on sunny days

0.1060° and 0.1287° respectively, which are lower than MS-PCA. The MS-PCA image denoising method proposed in this book can effectively reduce the AoP image noise while retaining useful solar meridian information, which significantly improves the orientation accuracy of the bioinspired polarization compass.

(2) Test and verification in the harsh weather environment of sand and dust

The test was conducted on the platform of the campus engineering training center building on March 16, 2023 (10:00–12:30). The test system is replaced by a domestic SLM-6S UAV with a three-axis turntable used in the sunny day test. The rest test system is the same as the sunny day test system. The entire test device is shown as Fig. 3.8.

Similarly, the BEMD decomposition method is used to adaptively decompose the original AoP image of sand and dust weather acquired by the polarization imaging system in the above test system into 10 BIMFs (i.e., BIMF1, BIMF2, BIMF3, …, BIMF10) and a residual component. The decomposition results of the original AoP image are shown as Fig. 3.9. Obviously, BIMF1 contains the most abundant solar meridian information and the solar meridian information contained in BIMF2 to BIMF10 gradually decreases.

Table 3.8 Comparison results of the heading error indexes of the static test on sunny days

Algorithm	Mean/°	SD/°	RMSE/°	UA/°
Raw data	0.1272	0.1837	0.2193	0.0034
BM3D	0.1115	0.1590	0.1933	0.0039
PDRDN	0.0963	0.1374	0.1642	0.0044
BM3D-KSVD	0.0811	0.1226	0.1463	0.0051
MS-PCA	0.0730	0.1060	0.1287	0.0058

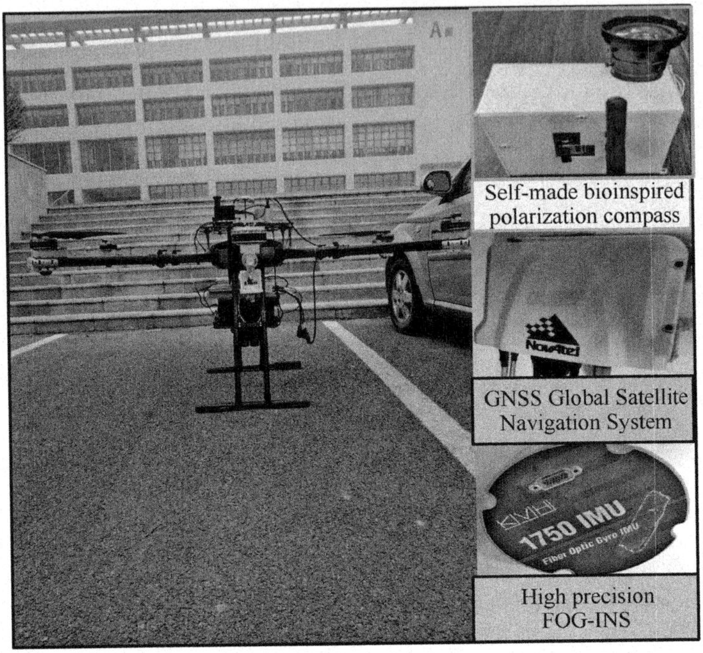

Fig. 3.8 Test map on sand and dust severe days

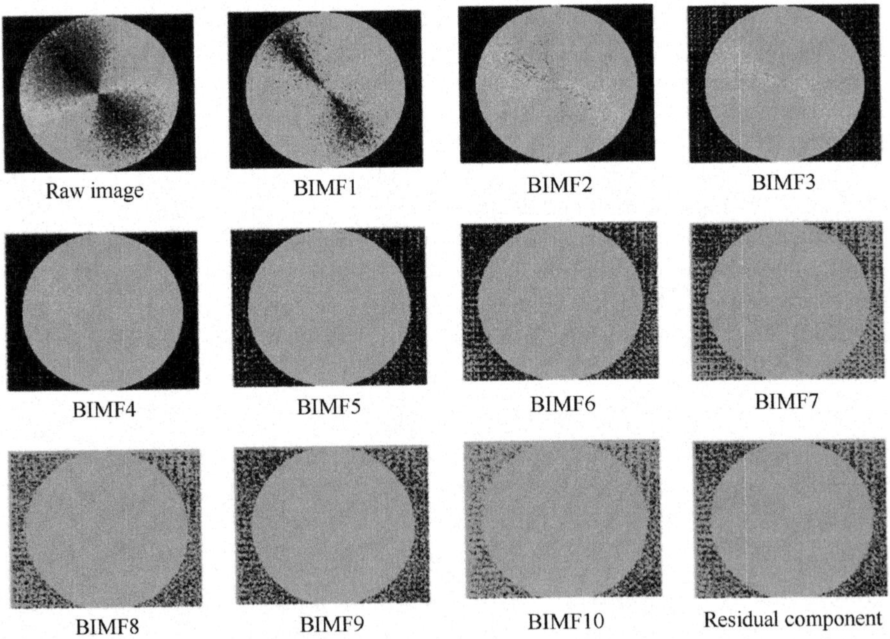

Fig. 3.9 Original and BEMD images in sand and dust days

The 1D-IE values for each BIMF and residual component after decomposition are calculated and the results are shown in Table 3.9. The decomposed BIMF and residual components are divided into S-BIMFs and NS-BIMFs by using the similarity of 1D-IE intervals. Finally, the PCA image denoising method based on principal component contribution rate is used to denoise the each BIMF scale of S-BIMFs and the NS-BIMFs are removed. The complete parameter settings of the MS-PCA denoising method are shown in Table 3.10. The denoising results of the original polarization angle image in sand and dust weather are shown as Fig. 3.10.

Table 3.11 shows the comparison results of the MS-PCA polarization angle image denoising method proposed in this book with the MG and SD after denoising the four original polarization angle images at 0°, 45°, 90° and 135° acquired under sand and dust weather conditions by existing image denoising techniques (such as BM3D, PDRDN and BM3D-KSVD). As can be seen from the table, the MS-PCA image denoising algorithm proposed in this study shows the best performance in image denoising.

In addition, Fig. 3.11 shows the comparison test results of the heading angle error calculated after removing the AoP image noise from the original AoP image acquired by the above-mentioned static test test and according to the proposed MS-PCA and other image denoising methods. It can be found from the heading angle error curves in the figure that the MS-PCA image denoising method proposed in this book finally effectively reduces the heading error caused by sand and dust weather.

Table 3.12 compares the mean (Mean), standard deviation (SD) and root mean square error (RMSE) of the static heading angle data obtained by various image denoising algorithms. It can be seen from the indicators in the table that the RMSE calculated by the MS-PCA method proposed in this book is 2.0407°, which is lower than the results obtained by other existing classical image denoising methods. All of above further prove that the MS-PCA method can effectively reduce AoP image noise while retaining useful solar meridian information, ultimately providing an effective method for orientation accuracy of the polarized light compasses in harsh weather environments such as sand and dust.

In summary, the MS-PCA image denoising algorithm based on multi-scale transformation proposed can effectively remove the noise in AoP images. It shows the best effect compared with other image denoising methods. For this reason, it plays a supporting role in improving the accuracy of the subsequent solution for the heading angle data.

Table 3.9 1D-IE of each BIMF and residual component after BEMD in sand and dust days

	BIMF1	BIMF2	BIMF3	BIMF4	BIMF5	BIMF6	BIMF7	BIMF8	BIMF9	BIMF10
1D IE	0.7270	1.3520	0.5997	0.9140	0.3964	0.5875	0.2753	0.3781	0.2291	0.2905

Table 3.10 Parameters of the denoising algorithm based on MS-PCA in sand and dust days

MS-PCA parameter	BIMF1		BIMF2		BIMF3		BIMF4		BIMF5		BIMF6	
	λ_1	R_{p1}	λ_2	R_{p2}	λ_3	R_{p3}	λ_4	R_{p4}	λ_5	R_{p5}	λ_6	R_{p6}
Numeric value	0.1280	0.1364	0.1160	0.1236	0.1120	0.1194	0.0990	0.1055	0.0900	0.0959	0.0750	0.0799

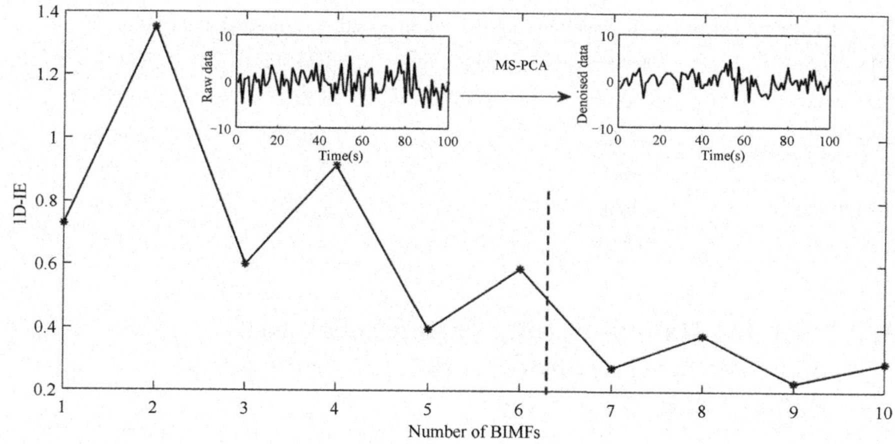

Fig. 3.10 Classification and results of BIMFs by MS-PCA

Table 3.11 Comparison results after various denoising algorithms in sand and dust days

Original image	Index	Raw data	BM3D	PDRDN	BM3D-KSVD	MS-PCA
0°	SD	23.7467	20.6737	20.6408	20.5672	16.9320
	MG	0.3092	0.3982	1.0752	1.1839	1.8699
45°	SD	23.7466	20.6736	20.6406	20.5674	17.0024
	MG	0.3091	0.3974	1.0743	1.1828	1.8774
90°	SD	23.7469	20.6732	20.6402	20.5675	16.9720
	MG	0.3089	0.3990	1.0760	1.1841	1.9113
135°	SD	23.7463	20.6734	20.6404	20.5674	16.9109
	MG	0.3088	0.3980	1.0747	1.1833	1.8753

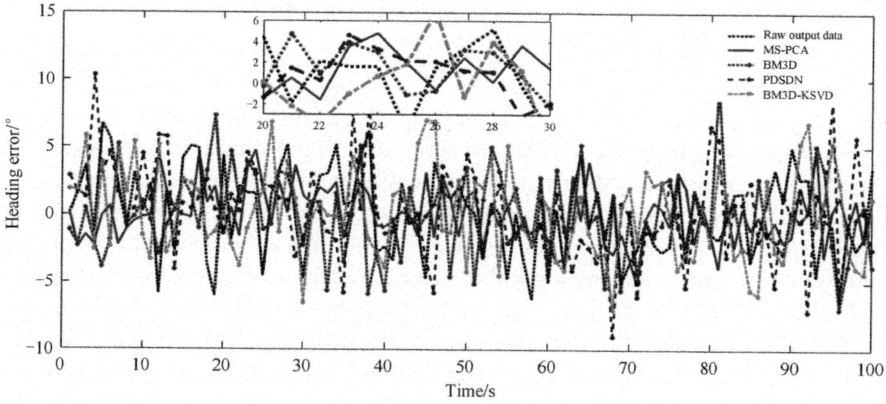

Fig. 3.11 Heading error of the static test in sand and dust days

Table 3.12 Comparison of the heading error based on a static test in sand and dust days

Algorithm	Mean/°	SD/°	RMSE/°	UA/°
Raw data	0.1921	3.2577	3.2485	0.3290
BM3D	0.0965	3.2577	3.0460	0.3439
PDRDN	0.0641	3.2266	2.8365	0.3050
BM3D-KSVD	0.0610	3.1218	2.7075	0.2058
MS-PCA	0.0104	3.0377	2.0407	0.3164

3.3 Heading Data Denoising Technology Based on Multi-Scale Transformation for Bioinspired Polarization Compass

According to the original heading angle data output from the FPGA hardware circuit after the denoised AoP image, it can be seen from the analysis of the noise generation mechanism and characteristics of the heading angle data in Sect. 3.1.2 that the noise generated by the FPGA circuit will have a non-negligible impact on the heading angle data. Therefore, in order to further improve the orientation accuracy of the bioinspired polarization compass, a novel heading data denoising method based on multi-scale transform (MST) is proposed. In this method, the Ensemble Empirical Mode Decomposition (EEMD) method is used to decompose the raw heading noise data output from the compass into different Intrinsic Mode Functions (IMF). The corresponding peak filtering technique (TFPF) is used to adaptively remove the noise in the original heading angle data.

3.3.1 Heading Data Denoising Technology Based on Multi-Scale Transformation

The promising heading data denoising technique based on Multiscale Transformation (MST) is based on the Ensemble Empirical Mode Decomposition (EEMD) method to superimpose the normally distributed white noise on the original heading angle data. It is adaptively decomposed into multiple different data Intrinsic Mode Functions (IMFs), and then its average value is taken as the final decomposition result. The corresponding data denoising algorithm is used to denoise the heading angle data.

In 2009, Membe et al. [22] proposed a noise-assisted signal analysis method, Ensemble Empirical Mode Decomposition (EEMD), to address the modal mixing phenomenon after directly decomposing data by empirical pattern (EMD). In this method, normally distributed white noise is introduced into each set of raw data so as to change the uniform distribution of the data series in different time–frequency spaces. The EEMD decomposition method is applied to decompose the data into multiple IMFs and the final average eigenmodal function ($\overline{IMF}$s) is obtained by calculating the average value of this group of IMFs. The normal distribution of white noise is used to achieve zero mean characteristics so that the real data signal can be well preserved. EEMD is also an adaptive signal multi-scale analysis method that decomposes nonlinear and nonstationary signals into multi-frequency IMFs and residual components (Residual) based on EMD and driven by a given input signal itself. Similar to the EMD decomposition results, the different components $\overline{\text{IMF}}$ of EEMD averaged by each group of IMFs also represent the specific details of the data signal at different scales. The residual component represents the overall trend of the signal. The low-order $\overline{\text{IMF}}$s suggests the high-frequency noise pattern and the high-order $\overline{\text{IMF}}$s hints the low-frequency signal pattern. The components $\overline{\text{IMF}}$s are locally orthogonal and able to effectively reflect the original signal feature information. Therefore, the algorithm has been widely used in the field of climate, engineering, and so on.

The steps to perform EEMD on a given noise data signal $S(t)$ are as follows [23]:

Step 1: Add normally distributed white Gaussian noise $N_M(t)$ to the original data signal $S(t)$ to get a new data signal $S_M(t)$.

$$S_M(t) = S(t) + N_M(t) \tag{3.18}$$

Step 2: Take the data signal with Gaussian white noise $S_M(t)$ as a whole, using EMD to decompose multiple IMF components, namely IMF_{11}, IMF_{12}, ..., IMF_{1L}.

Step 3: Determine the number of additions M.

$$M = (c_a/d_s)^2 \tag{3.19}$$

where c_a implies the amplitude coefficient of $N_M(t)$ and d_s is the standard deviation of the raw data signal and the sum of all IMFs obtained by decomposition.

Step 4: Repeat Step 1 and Step 2. After adding M new Gaussian white noise with normal distribution, IMF components are obtained through EMD, namely IMF_{21}, IMF_{22}, ..., IMF_{2L}, IMF_{M1}, IMF_{M2}, ..., IMF_{ML}.

Step 5: Average the IMFs obtained by decomposition after each addition of Gaussian white noise to get the final $\overline{\text{IMF}}_L$, i.e.

$$\overline{\text{IMF}}_L = \frac{1}{M}\sum_{l=1}^{M}\text{IMF}_{lL} = \frac{1}{M}(\text{IMF}_{1L} + \text{IMF}_{2L} + \ldots\ldots + \text{IMF}_{ML}) \tag{3.20}$$

The result of the decomposition of the entire given noise data signal $S(t)$ can be expressed as:

$$S_M(t) = \sum_{l=1}^{L}\overline{\text{IMF}}_l(t) + r_l(t) \tag{3.21}$$

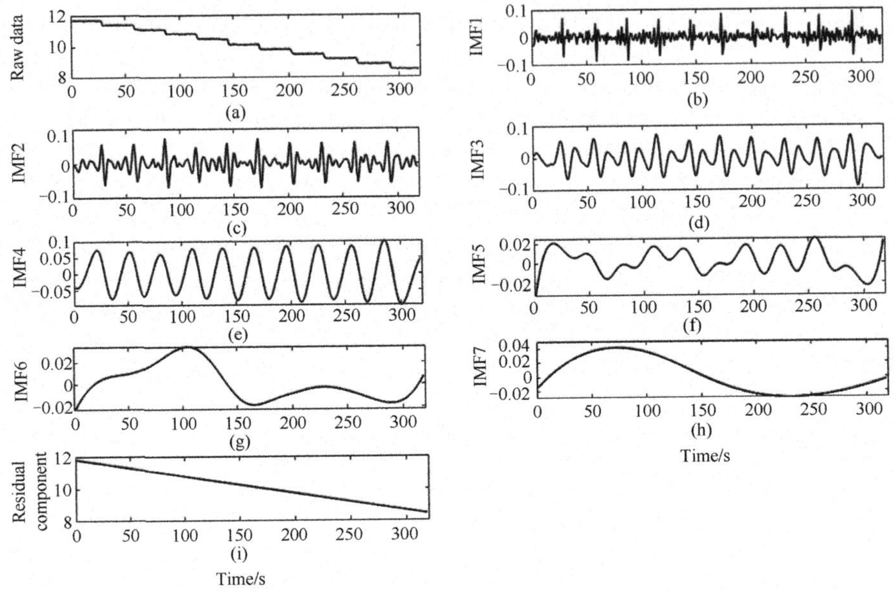

Fig. 3.12 EEMD diagram of the original heading angle data

where $\overline{\text{IMF}}_l(t)$ is the lth $\overline{\text{IMF}}s$ component obtained by EEMD decomposition, which L is the total number of $\overline{\text{IMF}}s$ components and $r_l(t)$ represents the residual component.

Taking the heading data studied in this book as an example, the whole decomposition of EEMD is shown as Fig. 3.12. Figure 3.12a is the source data signal. Figure 3.12b–h is the high frequency to low frequency and Fig. 3.12i is the residual component.

3.3.2 MS-TFPF Heading Data Denoising Technology Based on EEMD for Bioinspired Polarization Compass

From the analysis in the previous section, it can be seen that the new heading data denoising method based on Multi-Scale Transform (MST) can add normally distributed Gaussian white noise to the original heading angle data. EEMD adaptively decomposes the noisy data with normally distributed Gaussian white noise into multiple IMFs with high to low frequencies to obtain the final mean of the eigenpattern function $\overline{\text{IMF}}$s and residual components. In this section, a multi-scale time–frequency peak filtering (MS-TFPF) heading data denoising method based on EEMD decomposition is introduced for the original noisy heading angle data.

Time–Frequency Peak Filtering (TFPF) is an instantaneous frequency estimation method based on the Wigner-Ville distribution, in which the original noisy data signal is encoded into the instantaneous frequency (IF) of the analytic signal through frequency modulation. The instantaneous frequency (IF) is estimated by using the peak value of the Wigner-Ville distribution for the analytic signal as the effective signal estimation [24]. The denoising effect of TFPF with different window length is different. It is found that long-window TFPF is more suitable for high-frequency signals and short-window TFPF is more suitable for low-frequency signals [25]. Due to its strong superiority in data denoising, TFPF algorithm is widely used in seismic survey, fault diagnosis, and denoising of motor vehicle acceleration signals. Some improved TFPF-based algorithms, such as the TFPF-EWT algorithm [26], have also been proposed. Although the improved TFPF algorithm has a good denoising effect, the adaptive determination method of TFPF threshold parameter, namely, how to determine the window length still needs to be studied. In addition, the TFPF algorithm application in denoising the heading angle data for bioinspired polarization compass has been seldom reported. In this connection, this book further studies an improved TFPF heading data denoising method based on EEMD. The core contents of the method include: EEMD decomposition, $\overline{\text{IMF}}$s classification, and threshold selection of TFPF data denoising algorithm. The whole heading data denoising process is as follows:

Step 1: Add noise. The normally distributed Gaussian white noise sequence $N_M(t)$ is added to the original heading angle data $S(t)$ output from the polarization compass to obtain the heading angle data with white noise $S_M(t)$, namely:

$$S_M(t) = S(t) + N_M(t) \tag{3.22}$$

Step 2: Adopt EEMD to decompose the heading angle data $S_M(t)$.

The denoising method of MS-TFPF heading angle data based on EEMD firstly uses the EEMD data decomposition method introduced in Sect. 3.3.1 to decompose the heading angle data with normally distributed Gaussian white noise obtained by Step 1 into multiple IMFs with different frequencies and a residual component $r_l(t)$.The average value is calculated and ultimately achieves the ensemble mean $\overline{\text{IMF}}s$ of L intrinsic mode functions. M times of Step 1 and Step 2 are repeated and the final heading angle data is expressed as:

$$S_M(t) = \sum_{l=1}^{L} \overline{\text{IMF}}_l(t) + r_l(t) \tag{3.23}$$

Step 3: Classify the results of EEMD decomposition.

According to the need of this study, the sample entropy (SE), which can detect the similarity and complexity of time series, is adopted as the classification criterion to divide the L $\overline{\text{IMF}}$s series into low-frequency real signal components (LFT-C), mixed components (H-C) and high- frequency noise components (HFN-C) according to their frequencies from high to low.

Sample entropy (SE) is utilized to measure the similarity and complexity of the time series by the probability of generating a new pattern in the detected signal, regardless of the detected signal data length [27]. The lower SE value indicates the smaller probability of generating a new pattern and the higher self-similarity of the time series. In contrast, the higher SE value implies the greater probability of generating a new pattern and the lower self-similarity of the time series and the more complex the sample series. The sample entropy (SE) of each $\overline{\text{IMF}}$ is calculated as follows:

(1) A set of m-dimensional time vector sequences $\overline{\text{IMF}}_{m1}, \overline{\text{IMF}}_{m2}, \ldots, \overline{\text{IMF}}_{mL}$ consisting of L $\overline{\text{IMF}}s$ is obtained by averaging the above decomposition. For a given $\overline{\text{IMF}}_{mi}$ $(1 \leqslant i, j \leqslant L - m + 1, i \neq j)$, if the numbe $B_i^m(r)$ of $\overline{\text{IMF}}_{mj}$ vectors that satisfies $d\left[\overline{\text{IMF}}_{mi}, \overline{\text{IMF}}_{mj}\right] \leqslant r$ is counted, then:

$$B^m(r) = \frac{1}{L - m + 1} \sum_{i=1}^{L-m+1} B_i^m(r) \tag{3.24}$$

where $d\left[\overline{\text{IMF}}_{mi}, \overline{\text{IMF}}_{mj}\right]$ denotes the distance between $\overline{\text{IMF}}_{mi}$ and $\overline{\text{IMF}}_{mj}$. r suggests the similarity measure, which is the similarity tolerance namely.

(2) The M-dimensional time vector sequence is increased to N-dimensional and count to satisfy $d\left[\overline{\text{IMF}}_{m+1,i}, \overline{\text{IMF}}_{m+1,j}\right] \leqslant r$ the number $A_i^m(r)$ of vectors $\overline{\text{IMF}}_{mj}$, then:

$$A^m(r) = \frac{1}{L - m} \sum_{i=1}^{L-m} A_i^m(r) \tag{3.25}$$

(3) The sample entropy (SE) of L $\overline{\text{IMF}}s$ is:

$$\text{SE}(m, r, L) = -\ln\left[\frac{A^m(r)}{B^m(r)}\right] \tag{3.26}$$

Step 4: Select heading data denoising algorithm.

In order to effectively remove the noise in heading data, it is crucial to use the appropriate time–frequency spike filtering technique (TFPF) for each of the low-frequency true signal components (LFT-C) and mixed components (H-C $\overline{\text{IMF}}$). Using TFPF for each of all LFT-C and H-C $\overline{\text{IMF}}$ effectively attenuates noise but results in a large computational load, while using TFPF for a select few $\overline{\text{IMF}}$ of them reduces the computational load but has limited denoising effect. Therefore, in order to retain the real useful signal to the greatest extent and effectively reduce the noise, a new adaptive TFPF threshold selection method is proposed. In this method, the reference threshold (T_R) is used as the judgment criterion, which is gained before and after the removal of LFT-C noise by utilizing TFPF. If this threshold difference is greater than the preset threshold θ, the TFPF threshold for LFT-C noise removal is adjusted until the reference threshold (T_R) reaches the set threshold θ. Preset threshold θ is definited as [28]:

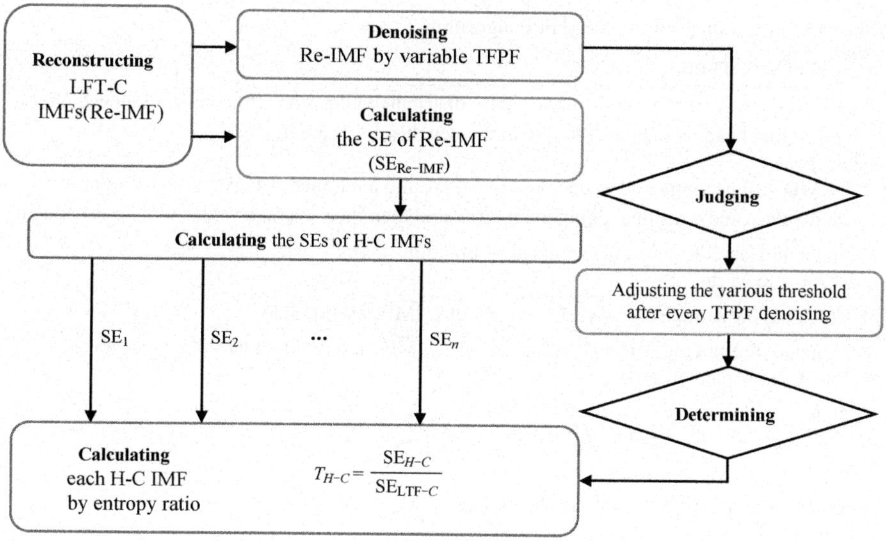

Fig. 3.13 Flow chart of the adaptive TFPF threshold selection method

$$\theta = \frac{\text{mean}\left\{\sum_1^q |u(i)|\right\}}{q} \tag{3.27}$$

where u represents the LFT-C vector and q is described as the length of u. The TFPF threshold (T_{H-C}) used to remove noise from the mixed component (H-C) is determined by the ratio of the average sample entropy (SE) for LFT-C and H-C to the reference threshold (T_R), namely (Fig. 3.13):

$$T_{H-C} = T_R \frac{\text{mean}(\text{SE}_{H-C})}{\text{mean}(\text{SE}_{LTF-C})} \tag{3.28}$$

Step 5: Denoise the noisy heading angle data.

According to the threshold parameters calculated in Step 4, the flexible characteristics of TFPF are fully utilized, that is, the short-window TFPF is used to remove the noise in the low-frequency real signal component (LFT-C) while maintaining the effective data as much as possible. The noise in the mixed component (H-C) is removed as much as possible by using the long-window TFPF, and the high-frequency noise component (HFN-C) is directly discarded finally.

Step 6: Reconstruct the heading angle data after denoising.

The denoised LFT-C and H-C are reconstructed to obtain the denoised heading data. The entire MS-TFPF algorithm flow is shown as Table 3.13 and Fig. 3.14.

Table 3.13 Flow chart of the MS-TFPF algorithm

Algorithm: MS-TFPF
Requirements: Add M times a subnormally distributed Gaussian white noise sequence $N_M(t)$ to the original heading data $S(t)$ to obtain the new heading data $S_M(t)$ **decompose** 1: **EEMD** will be employed to decompose $S_M(t)$ into a number of IMFs with different frequencies and a residual component $r_l(t)$, and then their average values will be calculated to obtain L the mean of the eigenpattern functions $\overline{\text{IMF}}$s 2:**for** $i = 1$ to M **do**
3: Calculate the sample entropy SE(m, r, L) of L $\overline{\text{IMF}}$s by Eq. 3.26
4: Compare the sample entropy SE(m, r, L) of L $\overline{\text{IMF}}$s and divide it into three categories: LFT-C, H-C, and HFN-C
5:end for 6: According to θ to calculate T_R and T_{H-C} of H-C **Denoise** T_{H-C}
7: Short-window TFPF is used to remove LFT-C noise, and long-window TFPF is used to remove H-C noise
8: Discard HFN−C directly **reconstruct**
9: Reconstruct the denoised LFT-C and H-C

3.4 Verification of Heading Data Denoising Based on Multi-Scale Transformation for Bioinspired Polarization Compass

In order to verify the effectiveness and practicability of the proposed data denoising technology based on multi-scale transformation for the bioinspired polarization compass, the turntable, static and UAV airborne test are carried out, respectively.

(1) Turntable test verification

The turntable test system is the same as the test system in Sect. 3.2.2. The outdoor turntable test was carried out on the campus on May 7, 2023 (15:30–17:00). The results are shown as Fig. 3.15. As can be seen from the graph, the true heading signal is completely submerged in the polarization compass noise analyzed in Sect. 3.1.2.

First, the heading angle data with 300 normally distributed white Gaussian noise (mean square error of 0.3) is decomposed by EEMD to obtain 7 $\overline{\text{IMF}}$s and one residual component, as shown in Fig. 3.16.

Table 3.14 shows that the sample entropy (SE) of each sample $\overline{\text{IMF}}$ is calculated from Eq. (3.26). $\overline{\text{IMF}}$s are divided into LFT-C, H-C, and HFN-C based on the classification principle of SE similarity.

The long-window TFPF denoises the H-C and the short-window TFPF denoises the LFT-C and retains the effective signal, directly eliminating the HFN-C. Finally, the denoised H-C and LFT-C are reconstructed to obtain the final denoised heading

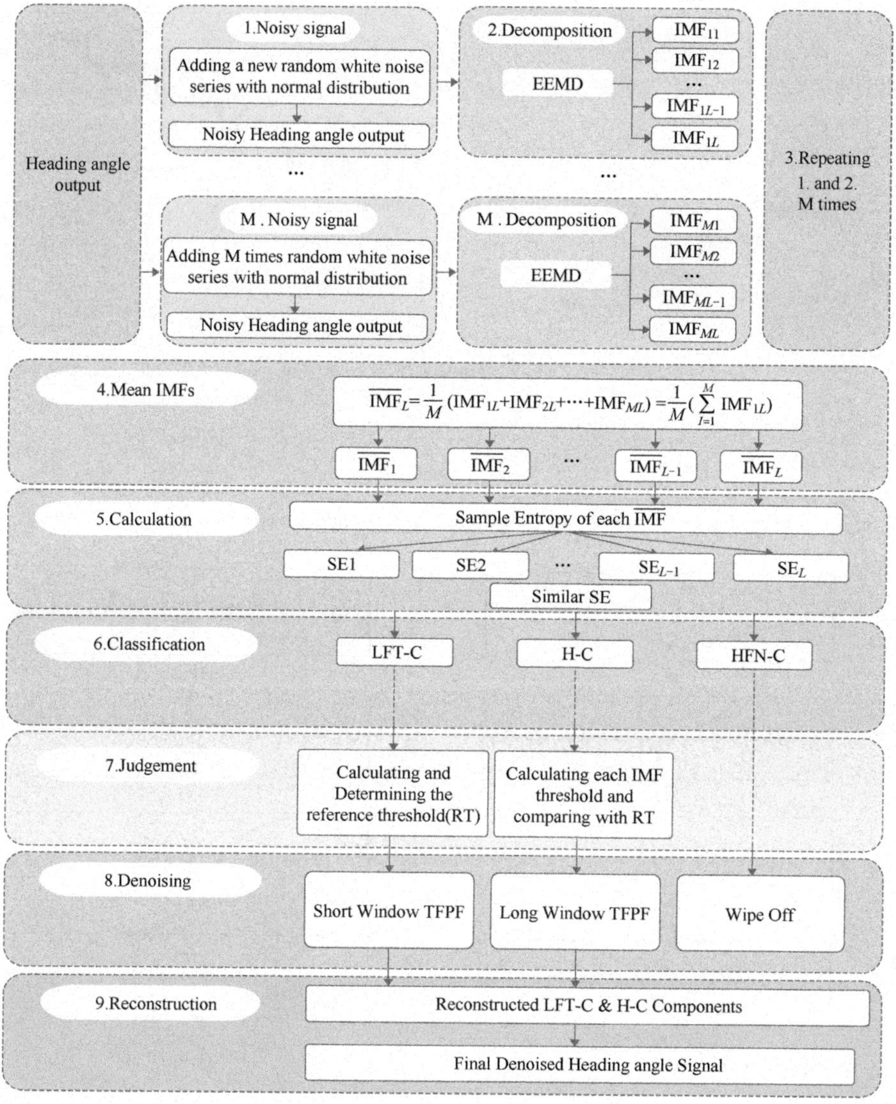

Fig. 3.14 Flow chart of the MS-TFPF algorithm

data signal.The results are shown in Fig. 3.17. The complete parameter settings of the MS-TFPF heading data denoising algorithm are shown in Table 3.15.

Figure 3.18a and b are the comparison results of heading angle and heading angle error obtained using turntable test to verify the MS-TFPF data denoising method proposed in this book and other classical data denoising methods (such as TFPF, RLowess and TFPF-EWT). Due to the inevitable spike noise data of heading angle during the turntable rotation, the adaptive TFPF algorithm is taken to denoise after

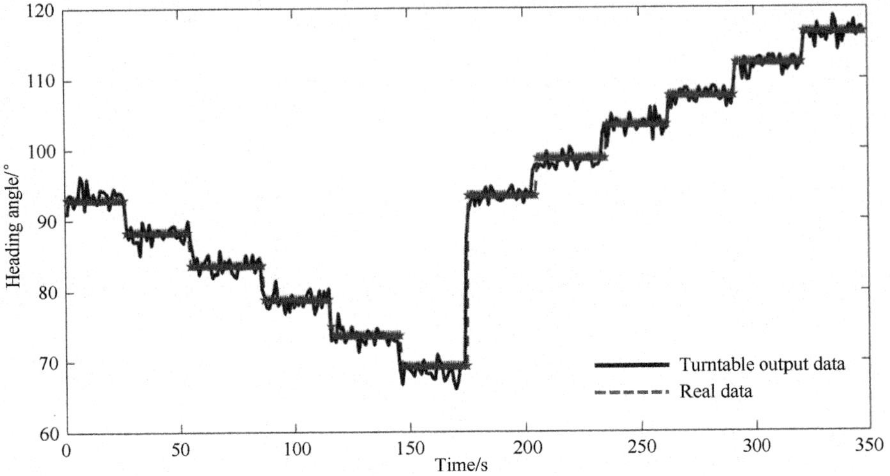

Fig. 3.15 Turntable test results of the bioinspired polarization compass

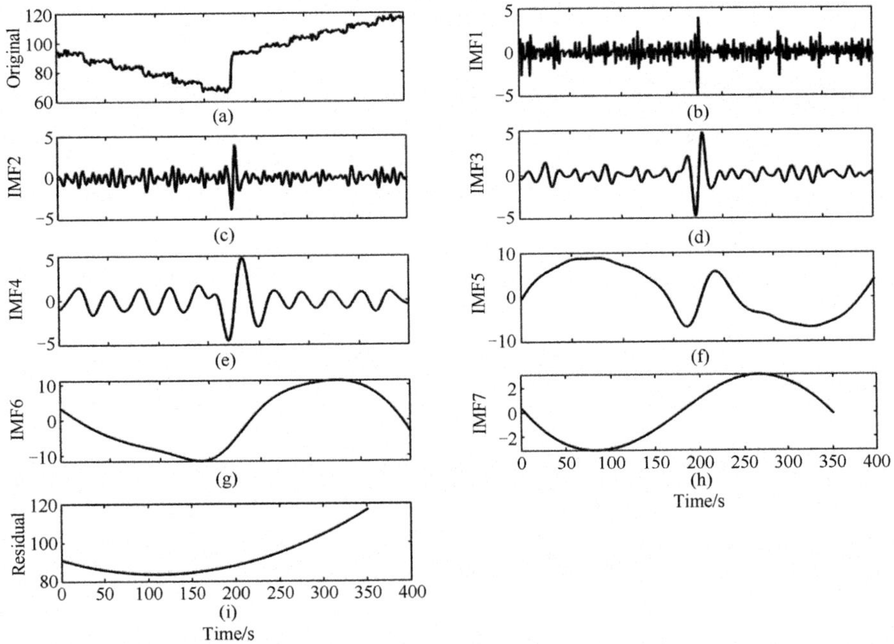

Fig. 3.16 Obtained $\overline{\text{IMF}}$s after EEMD using the turntable heading data

Table 3.14 Each $\overline{\text{IMF}}$s value calculated after EEMD

	$\overline{\text{IMF}}_1$	$\overline{\text{IMF}}_2$	$\overline{\text{IMF}}_3$	$\overline{\text{IMF}}_4$	$\overline{\text{IMF}}_5$	$\overline{\text{IMF}}_6$	$\overline{\text{IMF}}_7$
SE	1.5770	0.9393	0.5748	0.4240	0.1075	0.0529	0.0407

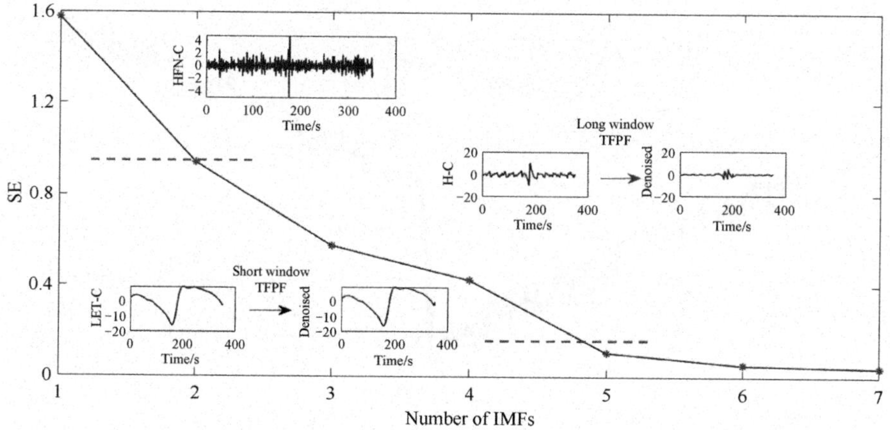

Fig. 3.17 IMFs classification with turntable test results and denoising by MS-TFPF

Table 3.15 Parameters of the heading data denoising algorithm based on MS-TFPF

	LFT-C IMFs				H−C IMFs	
MS-TFPF parameters	q	θ	RT	SELFT-C	TH−C	SEH-C
Parameter value	6000	2.3e−5	3.1	0.5520	12.2	2.1740

EEMD processing in the MS-TFPF algorithm. The spike noise data is also effectively suppressed and the heading error is greatly reduced.

Table 3.16 compares the Mean, SD, and RMSE of various heading angle data calculated from the denoised AoP image using the Sect. 3.2 image denoising algorithm. As can be seen from the table, the minimum Mean, SD and RMSE of the MS-TFPF algorithm are 0.0900°, 1.2383° and 1.2365° respectively, which indicats that the proposed algorithm is superior to other existing classical data denoising algorithms in improving the heading accuracy.

(2) Static test verification

The static test process is to keep the polarization compass installed on the three-axis turntable still to ensure that the output signal of the polarization compass is not affected by any movement. The heading angle data signal is finally output and calculated the heading angle error. The comparison results of the heading angle and heading angle error obtained by the MS-TFPF method proposed with other classical data denoising methods are shown as Fig. 3.19a and b. Using the same denoising process as the turntable test, the results display that the heading measurement error has a significant decreasing trend after the MS-TFPF data denoising.

In addition, Table 3.17 lists the Mean, SD, and RMSE of the static heading data signals obtained for each of the data denoising algorithms used for comparison. As can be seen from the table, the SD and RMSE of the MS-TFPF algorithm proposed

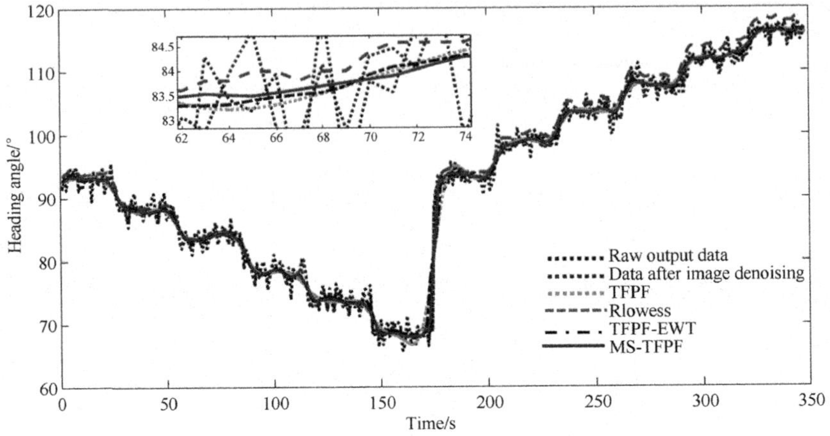

(a) Various data de-noising algorithm comparison in turntable test

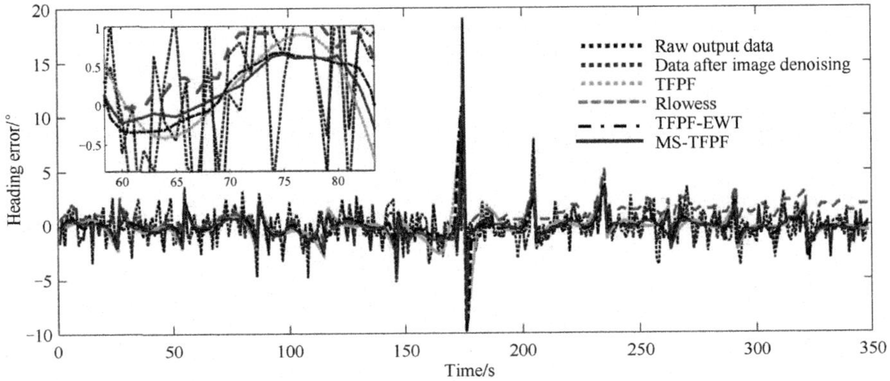

(b) Various data de-noising algorithms to obtain heading error

Fig. 3.18 **a** Various data de-noising algorithm comparison in turntable test, **b** Various data de-noising algorithms to obtain heading error

Table 3.16 Comparison of the heading error indexes obtained by the turntable test

Algorithm	Mean/°	SD/°	RMSE/°	UA/°
Raw data	0.1244	1.9164	1.9136	0.0665
Data after image denoising	0.1212	1.5315	1.5293	0.0743
TFPF	0.1106	1.5082	1.5060	0.0744
Rlowess	0.7853	1.3866	1.3846	0.0822
TFPF-EWT	0.0905	1.3856	1.3836	0.0809
MS-TFPF	0.0900	1.2383	1.2365	0.1029

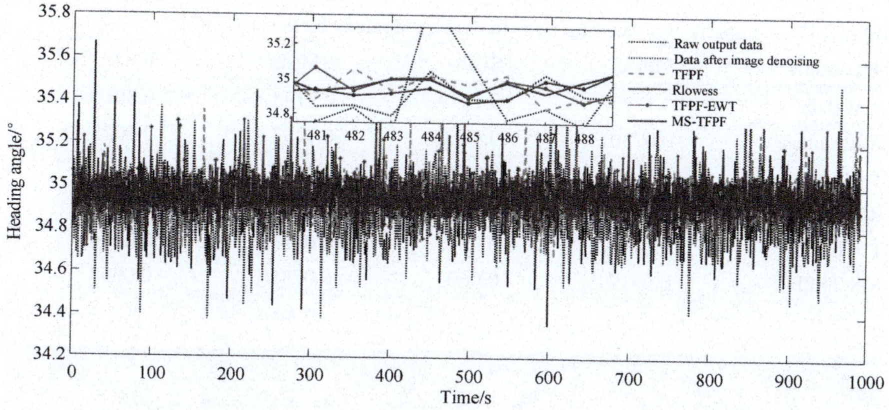

(a) Various data de-noising algorithms to obtain heading results in static test

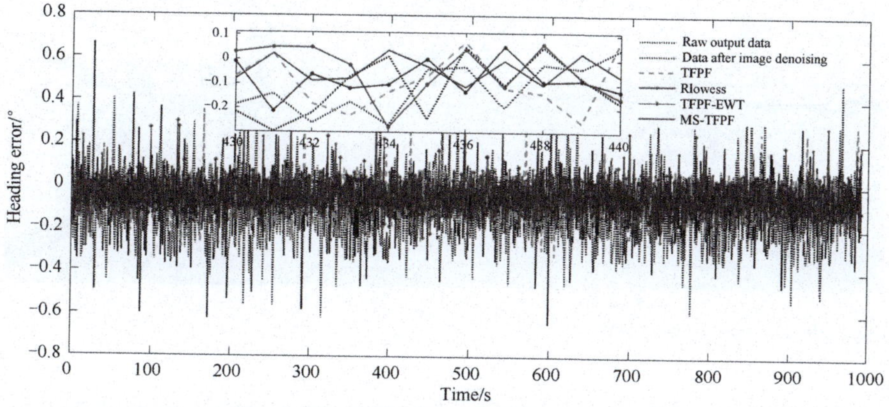

(b) Various data de-noising algorithms to obtain heading error results

Fig. 3.19 **a** Various data de-noising algorithms to obtain heading results in static test, **b** Various data de-noising algorithms to obtain heading error results

are 0.0605° and 0.0735° respectively, which further indicates that this data denoising algorithm has the best effect in reducing the heading angle error.

(3) UAV airborne test verification

The UAV airborne test is a flight test conducted around the campus engineering training center building. The test system is composed of a domestic SLM-6S UAV (maximum load of 15 kg, flight endurance of 20 min), the same self-made bioinspired polarization compass and FOG-INS/GNSS measurement reference system as the turntable test equipment, as shown in Fig. 3.20. The polarization imaging system in the polarization compass is used to directly collect the real polarization angle image.

Table 3.17 Comparison of heading error indicators obtained by the static test

Algorithm	Mean/°	SD/°	RMSE/°	UA/°
Raw data	0.1272	0.1837	0.2193	0.0019
After image	0.0730	0.1060	0.1287	0.0023
TFPF	0.0636	0.0923	0.1121	0.0026
Rlowess	0.0563	0.0817	0.0992	0.0029
TFPF-EWT	0.0490	0.0711	0.0863	0.0034
MS-TFPF	0.0417	0.0605	0.0735	0.0058

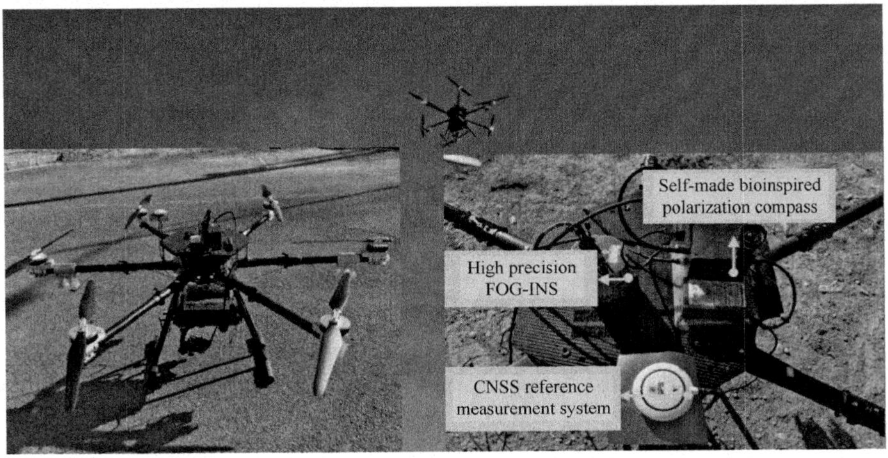

Fig. 3.20 Diagram of the UAV test

The heading angle data is calculated by the image denoising algorithm in Sect. 3.2 and directly output through the polarization compass.

The UAV flies at an altitude of 300 m to the ground and the flight distance is 500 m. Figure 3.21 displays the flight trajectory.

Figure 3.22a and b exhibit the results of the heading angle data obtained from the UAV airborne test and the comparison results of MS-TFPF with other classical data denoising methods. It can be seen from the figure that there are also spike data, which is caused by turning and shaking during the flight of the UAV. After using the MS-TFPF data denoising method, the generated spike noise is effectively suppressed and the whole heading error is significantly reduced, which suggests that the algorithm does have good performance in improving the UAV orientation accuracy.

Table 3.18 illustrates the Mean, SD, and RMSE for each data denoising algorithm in UAV flight test. As can be seen from the table, the SD and RMSE of the MS-TFPF data denoising algorithm proposed have the minimum values of 0.3118° and 0.3116° respectively, which further indicates that this data denoising algorithm is superior to

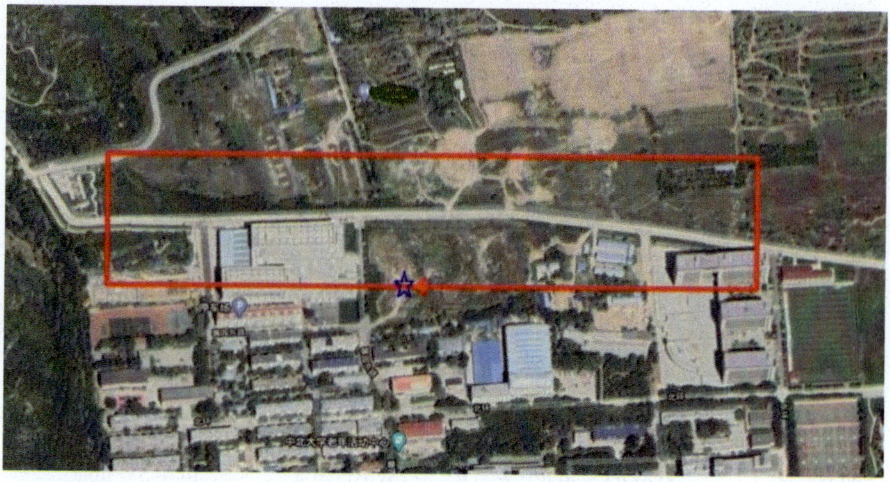

Fig. 3.21 Flight trajectory of the UAV test

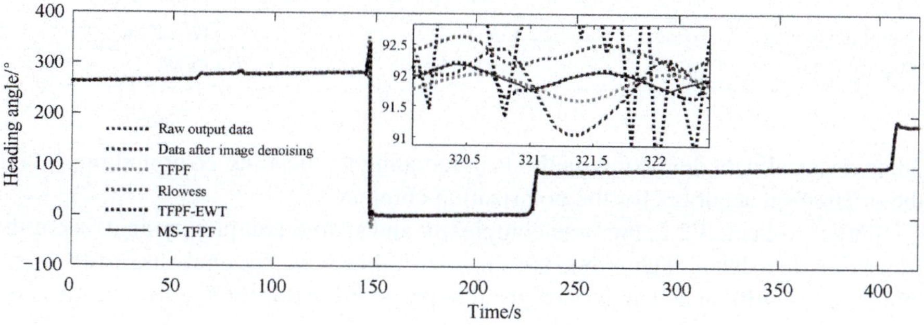

(a) Heading results obtained by various data denoising algorithms in UAV test;

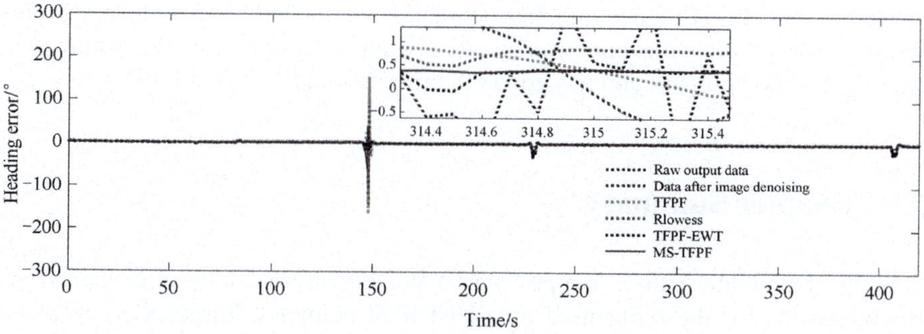

(b) Heading error results obtained by various data denoising algorithms

Fig. 3.22 **a** Heading results obtained by various data denoising algorithms in UAV test; **b** Heading error results obtained by various data denoising algorithms

Table 3.18 Comparison of heading error indicators obtained by the UAV test

Algorithm	Mean/°	SD/°	RMSE/°	UA/°
Raw data	0.894	1.0905	1.0900	0.1299
Data after image denoising	0.5154	0.7634	0.7631	0.1013
TFPF	0.3369	0.6386	0.6383	0.1133
Rlowess	0.1451	0.3369	0.3367	0.1230
TFPF-EWT	0.1458	0.3237	0.3235	0.1130
MS-TFPF	0.0944	0.3118	0.3116	0.1022

Table 3.19 Time complexity and space complexity based on MS-TFPF

Function	Time	Space
Initial value	2DEF$\cdot S$	[2 + 3S] float
Add noise	(3ADD + 4DEF + 1DIV + 1SUB) $\cdot S \cdot K$	[4 + 5S] float
Compute $\overline{\text{IMF}}$s	O(N^2)	[6 $K \cdot N \cdot S$+8S] float
Iterative training	4DEF + (3CMP + 2ADD + 1SUB) $\cdot N \cdot K \cdot S$	[4] float
Weighted average	(6SUB + 22DEF)$\cdot N \cdot K \cdot S$	[28] float
Complexity	O(N^2)	O(N)

other classical data denoising methods in reducing the heading error and improving the orientation accuracy for the polarization compass.

Similar to Sect. 3.2.2, the time complexity and spatial complexity analysis of the MS-TFPF data denoising method are shown in Table 3.19. The analysis results reveal that the MS-TFPF data denoising algorithm proposed in this book is also an effective polynomial time solvable algorithm.

From the above test results, it can be seen that the Mean, SD and RMSE of the heading error are the smallest after the MS-TFPF data denoising method, especially the spike noise in the UAV turning process can also be effectively suppressed. Therefore, the MS-TFPF heading data denoising method based on multi-scale transformation proposed in this book can effectively improve the orientation accuracy for the polarization compass and has good effectiveness and practicability.

3.5 Chapter Summary

This chapter emphasizes a comprehensive noise analysis of the mechanism and characteristics for the bionspired polarized light compass. Targeted research has been conducted on AoP image denoising methods and heading angle data denoising methods based on multi-scale transformation to effectively improve the orientation accuracy for the bioinspired polarized light compass.

First, according to the polarization imaging principle of polarization compass and the source of heading angle data, the generated mechanism and characteristics of AoP image noise and heading angle data noise are comprehensively analyzed. The results reveal that the polarization imaging system and output data noise from the bioinspired polarization compass are mainly yeilded by the photon fluctuation and the electronic fluctuation in the charge voltage conversion and subsequent electrical signal amplification in the pixel circuit, and finally the conversion between analog signal and digital signal.

Secondly, aiming at the AoP image noise, integrated with the two-dimensional empirical mode decomposition method (BEMD) and the feature information required by the research object, an AoP image denoising strategy based on multi-scale transformation (MS-PCA) is proposed. The core contents include: BEMD decomposition, BIMFs classification, and threshold selection of PCA image denoising algorithm. From the results of outdoor static tests conducted on sunny and dusty weather, it can be seen that the MS-PCA polarization angle image denoising method proposed in this chapter is very effective and practical in suppressing AoP image noise, providing strong assistance for subsequent heading angle calculation.

Finally, in view of the noisy heading angle data after the denoised AoP image is output from the compass, integrated with the ensemble empirical modal decomposition (EEMD) method and a time–frequency peak filtering (TFPF) algorithm, a heading data denoising strategy based on multi-scale transformation (MS-TFPF) based on multi-scale transformation is proposed. The core contents include EEMD decomposition, IMFs classification, and threshold selection of TFPF data denoising algorithm. Outdoor static, turntable and UAV airborne tests verify that the MS-TFPF data denoising algorithm proposed in this book has good results in further improving the orientation accuracy for the polarization compass.

Various experimental results demonstrate that the denoising method based on multi-scale transformation proposed in this book compared with the existing classical image and data denoising techniques can effectively improve the orientation accuracy for the polarization compass by suppressing the polarization compass image and data noise. With the development of hardware technology in the future, the denoising method based on multi-scale transformation proposed in this book can be transplanted to the polarization compass and be applied in practice.

References

1. Gow RD, Renshaw D, Findlater K, et al. A comprehensive tool for modeling CMOS image-sensor-noise performance. IEEE Trans Electr Dev. 2007;54(6):1321–9.
2. Donghua Z, Jun T, Xindong W, Chong S, Jun L. A multiscale transform denoising method of bionic polarization compass for improving the unmanned aerial vehicle navigation accuracy. Chin J Aeronaut. 2022;3(54):400–14.
3. Chengshuai Z, Xindong W, Donghua Z, Chenguang W, Xiaojie L, Chong S. Bionic polarization compass denoising method based on EMD-TFPF. Navig Position Timing. 2021;8(05):38–44.

4. Rudin LI, Osher S, Fatemi E. Nonlinear total variation based noise removal algorithms. Phys D. 1992;60(1–4):259–68.
5. Abergel R, Louchet C, Moisan L. Total variation restoration of images corrupted by poisson noise with iterated conditional expectations. New York: Springer; 2015. p. 178–90.
6. Buades A, Coll B, Morel JM. A review of image denoising algorithms with a new one. Multiscale Model. 2005;4(2):490–530.
7. Deledalle CA, Tupin F, Denis L. Poisson NL means: unsupervised non local means for Poisson noise. IEEE Int Conf Image Process. 2010;12:801–4.
8. Yu B, Lou L, Li S, Zhang Y, Qiu W, Wu X, Wang M, Tian B. Prediction of protein structural class for low-similarity sequences using Chou's pseudo amino acid composition and wavelet denoising. J Mol Graph Model. 2017;76:260–74.
9. Maggioni M, Katkovnik V, Egiazarian K, Foi A. Nonlocal transform-domain filter for Cubature data denoising and reconstruction. IEEE Trans Image Process. 2013;22(1):119–34.
10. Huang NE, Shen Z, Long SR, et al. The empirical pattern decomposition and the Hilbert spectrum for nonlinear and non-stationary time series analysis. Proc Roy Soc Lond A. 1998;454:903–94.
11. Perkins R, Gruev V. Signal-to-noise analysis of Stokes parameters in division of focal plane polarimeters. Opt Express. 2010;18(25):25815–24.
12. Graham M, Wei-Liang H, Alba P, et al. Liquid crystal polymer full-stokes division of focal plane polarimeter. Opt Express. 2012;20(25):27393–409.
13. Goudail F. Noise minimization and equalization for Stokes polarimeters in the presence of signal-dependent Poisson shot noise. Opt Lett. 2009;34(5):647–9.
14. Li XB, Li HY, Lin Y, et al. Learning-based denoising for polarimetric images. Opt Exp. 2020;28(11):16309–421.
15. Abubakar A, et al. A block-matching and 3-D filtering algorithm for gaussian noise in DoFP polarization images. IEEE Sens J. 2018;18(18):7429–34.
16. Ye WB, et al. A K times singular value decomposition based image denoising algorithm for DoFP polarization image sensors with Gaussian noise. IEEE Sens J. 2018;18(15):6138–44.
17. Abubakar A, Zhao XJ, Takruri M, et al. A hybrid denoising algorithm of BM3D and KSVD for Gaussian noise in DoFP polarization images. IEEE Access. 2020;8:57451–9.
18. Nunes JC, Bouaoune Y, Delechelle E, et al. Image analysis by bi-dimensional empirical pattern decomposition. Image Vis Comput. 2003;21(12):1019–25.
19. Guo S, Luan F, Song X, et al. Self-adaptive image denoising based on bidimensional empirical pattern decomposition (BEMD). Bio-Med Mater Eng. 2014;24(6):3215–22.
20. Yu MB, Wang B, et al. Transient power quality denoising based on EMD and PCA. Inform Sci. 2018;18:149–57.
21. Zhang L, Lukac R, Wu X, et al. PCA-based spatially adaptive denoising of CFA images for single-sensor digital cameras. IEEE Trans Image Process. 2009;18(4):797–812.
22. Membe YS, He Z, Li X, et al. An optimization-based ensemble EMD for classification of hyperspectral images. IEEE Int Instrum Measur Technol Confer. 2013;7(15):1–24.
23. Chen XY, Cui BB. Efficient modeling of fiber optic gyroscope drift using improved EEMD and extreme learning machine. Signal Process. 2016;128:1–7.
24. Liu YP, Li Y, et al. An amplitude-preserved time-frequency peak filtering based on empirical pattern decomposition for seismic random noise reduction. IEEE Geosci Remote Sens Lett. 2014;11(5):896–900.
25. Jun L, Donghua Z, Chenguang W, Jun T, Chong S. Attitude calculation method based on full-sky atmospheric polarization pattern. Rev Sci Instrum. 2019;90:015009.
26. Liu NH, Yang Y, Liu Z, et al. Seismic signal de-noising using time-frequency peak filtering based on empirical wavelet transform. Acta Geophys. 2020;68(2):425–34.
27. Xua MJ, Shanga PJ, Huang JJ. Modified generalized sample entropy and surrogate data analysis for stock markets. Commun Nonlinear Sci Numer Simul. 2016;35:17–24.
28. Xiaojie L, Xiaoting G, Donghua Z, Chong S, et al. INS/vision integrated navigation system based on a navigation cell model of the hippocampus. Appl Sci Basel. 2019;9(2):234.

Chapter 4
Orientation Error Modeling and Compensation Technology for Bioinspired Polarization Compass

The tilt will inevitably occur due to the carrier motion when the bioinspired polarization compass is attached to the mobile carrier (such as a drone), which brings about the zenith position in AoP images collected by the polarization imaging system to be changed. The error of fitting the solar meridian causes the AoP calculated error, decreasing the orientation accuracy for the bioinspired polarized compass dramatically. In addition, we have found in a large number of dynamic experiments that the coupling between the solar meridian and the carrier body axis angle (A-SMBA) and the inclination angle also produces very significant orientation errors. Therefore, this chapter first comprehensively analyzes the influence of the changed attitude angles by the polarization compass (including A-SMBA, pitch angle and roll angle) on the heading error. Next, the Gated Recurrent Unit (GRU) neural network is introduced to model and compensate for the orientation error caused by the changed attitude angles by the polarization compass so as to improve the orientation accuracy during the carrier actual motion.

4.1 Polarization Orientation Error Analysis and Model

4.1.1 Analysis of Polarization Orientation Error

In the practical application of a bioinspired polarization compass, the compass orientation accuracy will be significantly reduced if the simplified heading angle calculation method is adopted. Accordingt, it is essential to consider the compass tilt (i.e., pitch and roll) attached to a moving carrier and the formula for solving the heading angle is as follows [1]:

D. Zhao, *Intelligent Information Processing for Polarization Compass and Inertial Integrated Navigation System*, Navigation: Science and Technology 15,
https://doi.org/10.1007/978-981-97-7135-6_4

$$\varphi = \arcsin\left[\frac{-(\cot\phi\sin\delta+\sin\theta\cos\delta)\tan h_s}{\sqrt{(\cot\phi\cos\delta-\sin\theta\sin\delta)^2+\cos^2\theta}} - \arctan\frac{\cot\phi\cos\delta-\sin\theta\sin\delta}{\cos\theta} - \alpha_s\right] \tag{4.1}$$

$$\varphi = \pi - \arcsin\left[\frac{-(\cot\phi\sin\delta+\sin\theta\cos\delta)\tan h_s}{\sqrt{(\cot\phi\cos\delta-\sin\theta\sin\delta)^2+\cos^2\theta}} - \arctan\frac{\cot\phi\cos\delta-\sin\theta\sin\delta}{\cos\theta} - \alpha_s\right] \tag{4.2}$$

where φ, θ, δ and ϕ represent the heading angle, pitch angle, roll angle, and AoP, respectively. α_s and h_s imply the sun azimuth and altitude angles, respectively. Both of them can be achieved from the specific time and local location information from the astronomical ephemeris.

Equations 4.1 and 4.2 show that the polarization compass orientation error is closely related to the sun position (i.e., the solar azimuth α_s and solar altitude angle h_s), the AoP ϕ collected by the polarization imaging system, and the compass pitch θ and roll angle δ. Among them, the compass orientation error caused by natural factors such as the sun position (α_s and h_s) is not controlled by human factors. The AoP image ϕ collected by the polarization imaging system is the orientation error caused by the device factor, which can be solved by calibrating the device and by suppressing the noise in images and data with the proposed ways in Chap. 3. The carrier pitch and roll changes result to the zenith position in AoP images collected by the polarization imaging system to be changed. Correspondingly, the AoP calculated error is passed to the fitting of the solar meridian solar and solving heading angle. Eventually, the orientation accuracy for the bioinspired polarized compass is cut down dramatically. Figure 4.1a shows a schematic diagram of the change of orientation error with the roll angle δ when the pitch angle θ of the bioinspired polarization compass is fixed, and Fig. 4.1b shows a schematic diagram of the orientation error change with the pitch angle θ when the bioinspired polarization compass roll angle δ is fixed. It can be seen that there is a highly complex nonlinear relationship between the carrier pitch and roll changes of and the orientation error of the bioinspired polarization compass.

In addition, a large number of dynamic experiments have manifested that the coupling between the angle between the solar meridian and the carrier axis (A-SMBA) and the tilt angles of the bioinspired polarization compass (i.e., the pitch angle θ and roll angle δ) will produce a more significant orientation error. Namely, the orientation error of the polarization compass is not only related to its tilt angles, but also closely related to the coupling of tilt angles and A-SMBA. In order to study the influence of the coupling effect on the orientation error for the bioinspired polarization compass, as shown in Fig. 4.2a, the turntable roll angle δ is fixed at 0° and the pitch angle θ is changed from 6° to −6°. Each curve in the figure shows the trend of the polarization compass heading error with A-SMBA from 0° to 360°. The test results in Fig. 4.2b are the same as Fig. 4.2a but the turntable roll angle δ is fixed at 3°. Figure 4.2c shows that the pitch angle θ of the turntable is fixed at 0° and the roll angle δ changes from 6 to −6°. Each curve in the figure shows the trend of the heading error of the polarization compass with A-SMBA from 0° to 360°. Figure 4.2d is the same as Fig. 4.2c but the pitch angle θ of the turntable is fixed at 3°. It is clear from the above four diagrams: (1) The coupling effect between A-SMBA,

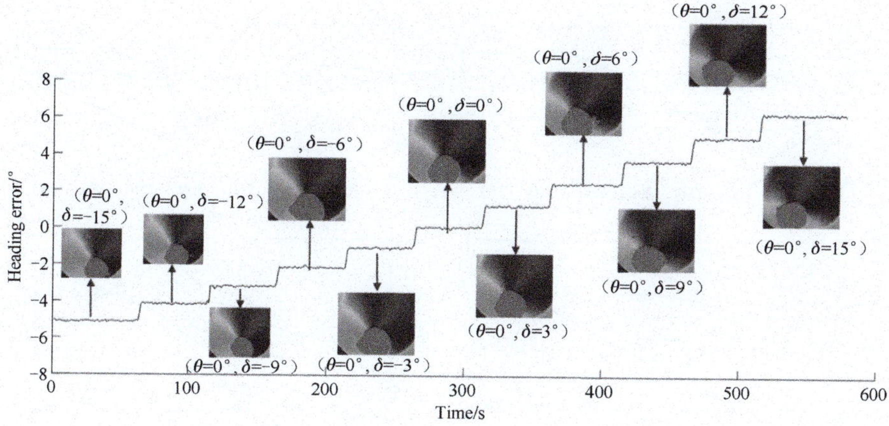

(a) Experimental result of the heading error varying with roll angle δ

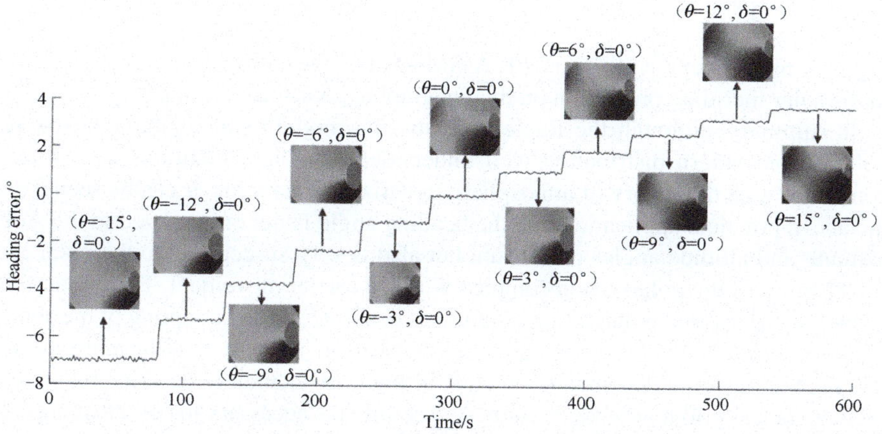

(b) Experimental result of the heading error varying with pitch angle θ

Fig. 4.1 **a** Experimental result of the heading error varying with roll angle δ, **b** Experimental result of the heading error varying with pitch angle θ

pitch angle θ, and roll angle δ has a significant impact on the orientation error of the polarized light compass; (2) Not only the tilt angles of the bioinspired polarized compass has a regular effect on the orientation error, but also A-SMBA has a regular influence on the orientation accuracy with the change of the relative position between the compass and the sun; (3) The orientation error curve of each graph in Fig. 4.2 has an intersection point and the two adjacent intersection points are separated by 180°. At each intersection, the orientation error error remains unchanged regardless of the change of inclination. The intersection points in Fig. 4.2a and b occur at 0°, 180°, and 360° A-SMBA. Namely, the carrier axis is parallel to the solar meridian projection on the geographic coordinate system. Similarly, the intersection points in

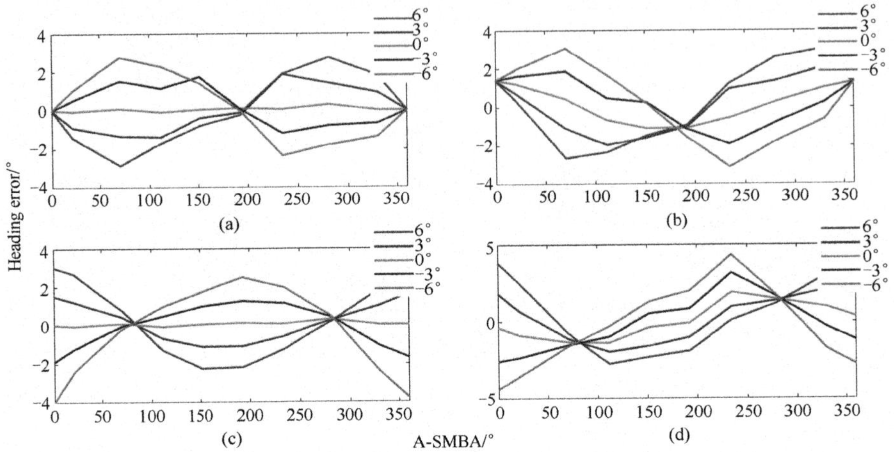

Fig. 4.2 2D variation schematic diagram of the heading error with A-SMBA and tilt angles

Fig. 4.2c and d occur at 90° and 270° A-SMBA, i.e., the carrier axis is perpendicular to the solar meridian projection on the geographic coordinate system.

In summary, although the Rayleigh scattering model explains the atmospheric polarization pattern distribution well under ideal weather conditions and is very consistent with the theory of atmospheric polarization patterns, it can be seen from the above practical application that the heading angle is not only related to the AoP, azimuth and altitude angles of the sun, but also is very susceptible to the pitch and roll changes of the polarization compass with the carrier movement. The orientation error of a bioinspired polarization compass is generated by the coupling of the above factors. Additionally, the above mapping relationship is highly nonlinear, uncertain, and long-series time dependent, which is difficult to accurately describe by simple models such as polynomial regression. As a result, it is necessary to take advantage of the deep learning neural network to effectively model and compensate the orientation error caused by the changed attitude angles of a bioinspired polarization compass in order to improve the orientation accuracy for the bioinspired polarization compass furtherly.

4.1.2 Model Construction for Polarization Orientation Error

The analysis in the previous section shows that the orientation error of a bioinspired polarization compass is closely related to the change of compass attitude angles. Therefore, in order to reduce the influence of polarization compass attitude angle changes and the coupling between them on the orientation accuracy, a method based on deep learning neural network is proposed to model and compensate the orientation error caused by the attitude angle change of bioinspired polarization compass in this book. The complex nonlinear relationship between the pitch angle θ, roll angle δ, A-SMBA and the heading error is established by training the neural network. The compass orientation accuracy is finally improved by compensating for the heading error.

In recent years, machine learning neural network models have been flourished and they are widely used in modeling nonlinear data and systems thanks to a general and flexible function approximation method. Neural networks such as BP neural networks, Elman neural networks, radial basis function (RBF) neural networks, and random forest (RF) can all be used for modeling nonlinear data and their effectiveness in modeling nonlinear systems has been demonstrated. However, it is difficult to determine the optimal time step for a model associated with a long series of data using a machine learning model and obtain the best prediction accuracywithout any prior knowledge.

With the vigorous development of deep learning neural networks, such as recurrent neural networks (RNNs), one of the most popular neural networks, they have achieved great success in many fields such as computer vision and natural language processing due to their ability to process long sequence data [2]. Although RNNs have proven to be an effective tool for processing time series prediction data, traditional RNNs have problems such as gradient vanishing and gradient explosion. Improved RNNs, such as long-short-term memory (LSTM) neural networks, can solve this problem well by introducing additional unit states [3]. In addition, the Gated Recurrent Unit (GRU) neural network that is another simplified model of LSTM exhibits comparable performance to LSTM in processing many time series data owing to a simpler structure and faster computation speed compared to LSTM [4]. Although GRU neural networks have been successful in many fields, they have not yet been applied in modeling and compensating the orientation error for the bioinspired polarization compass. Therefore, the GRU deep learning neural network is exploited to model and compensate for the the attitude angle change of the polarization compass fixed to the carrier in this book.

In the process of establishing the orientation error model caused by the change of attitude angle of the bioinspired polarization compass with GRU deep learning neural network, the current moment A-SMBA cannot be used as the input of the network model because the accurate A-SMBA of the current moment cannot be obtained without heading error compensation. Considering that the compass sampling frequency is high (30 Hz) in the actual orientation process, the A-SMBA changes between the previous moment and the current moment is small and the previous moment A-SMBA (i.e., PA-SMBA) is applied instead of the current moment A-SMBA as one of the model input vectors. Namely, PA-SMBA α, pitch angle θ and roll angle δ are regared as the input vector of the model and the orientation error is used as the model output vector directly. The structure is shown as Fig. 4.3.

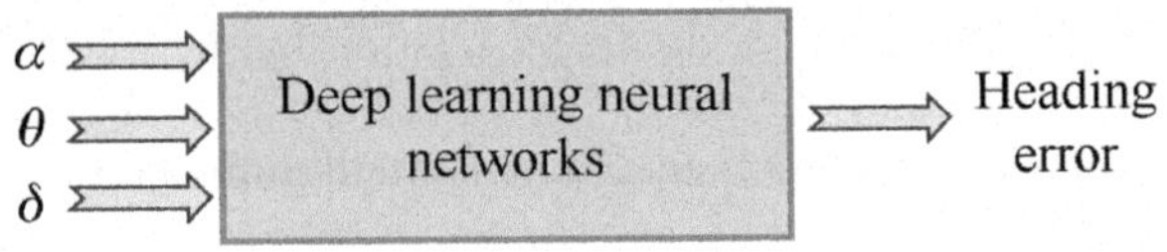

Fig. 4.3 Schematic diagram of the orientation error modelled by deep learning neural networks

4.2 Typical Neural Network Models

4.2.1 Recurrent Neural Network (RNN)

Recurrent Neural Network (RNN) was first proposed by Salinas et al. [5]. It is a neural network composed of a series of identical basic structural modules that can be used to process sequence data. All model parameters in the basic structural module can be shared at various positions in the sequence data and all loop units are connected in a chain structure. RNN includes input layer, hidden layer, and output layer. The biggest difference between RNN and other traditional artificial neural networks is that RNN has memory function. The hidden layer of this neural network is related to both the current network input and the output of the previous hidden layer.

The structure of the RNN neural network is shown as Fig. 4.4, in which t represents the time step in order to reconstruct the input data time series into a neural network learning dataset. The input dataset is usually preprocessed by changing the window size as a variable, which is called the time step [6]. n_t is the input vector of the time step. s_t suggests the hidden state of the neural network at the time step t. It is calculated by the hidden state s_{t-1} of the previous time step $t-1$ and the input vector n_t of the current time step t. h_t is the output vector of the time step t.

For an input time series $N = (n_1, n_2, \ldots, n_t)$ with a time step t, each column vector of $N = (n_1, n_2, \ldots, n_t)$ is input into the RNN neural network in sequence. The output h_t operation process is as follows through the RNN network model for

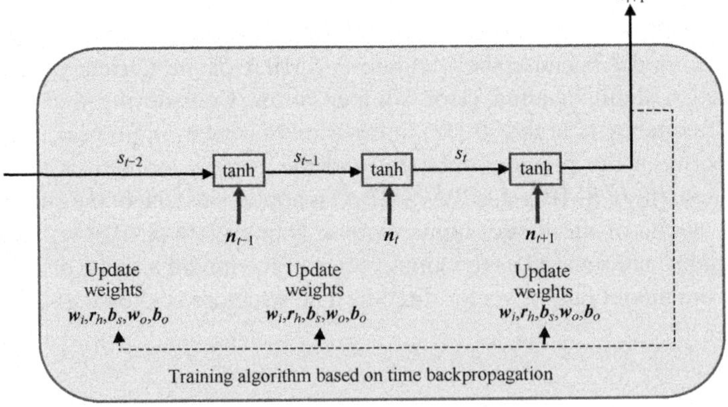

Fig. 4.4 Principle structure diagram of the RNN neural network

the input n_t at a certain time:

$$s_t = \tanh(w_i n_t + r_h s_{t-1} + b_s) \quad (4.3)$$

$$h(t) = \sigma(w_o s_t + b_o) \quad (4.4)$$

where tanh is the hyperbolic tangent activation function of the recurrent unit of the RNN neural network. The trainable w_i, r_h, w_o, b_s, and b_o represent the connection weight matrix of the input layer and the hidden layer of the neural network, the value of the hidden layer at the previous time as the weight matrix of the input at the current moment, the weight matrix and the bias matrix of the loop between the hidden and output layers, respectively. $\sigma(\cdot)$ is the softmax activation function of the neural network. For each time step, the corresponding basic structure has the same parameters w_i, r_h, w_o, b_s, and b_o.

The RNN neural network employs the temporal backpropagation training algorithm (BPTT) [7] to find the derivatives of all the connection weight matrices and bias matrices for the model, continuously updating them until the neural network model converges. In this process, the stochastic gradient descent (SGD) method of logarithmic loss function (L) is introduced to continuously modify the parameters of the RNN model to minimize the LF. That is to say that the negative gradient of the training parameters to be optimized is continuously searched and updated until the neural network model converges. This process consists of the following three steps:

Step 1: Calculate the output value of each neuron in the RNN neural network forward using Eqs. (4.3) and (4.4).

Step 2: Calculate the final loss function. Due to the fact that the sequence has a loss function $L(t)$ at each time step in RNN neural networks, the final loss function L is the sum of the loss functions at each time step:

$$L = \sum_{j=1}^{t} L(j) \quad (4.5)$$

Step 3: Calculate the parameters w_i, r_h, w_o, b_s and b_o. The gradient ∇_t of the hidden state s_t at the current time t is calculated by differentiation, and then the gradients of the corresponding parameters w_i, r_h, w_o, b_s and b_o are calculated. The entire process is expressed mathematically as:

$$\nabla_t = \frac{\partial L}{\partial s_t} \quad (4.6)$$

$$\nabla w_i = \frac{\partial L}{\partial w_i} = \sum_{j=1}^{l} \frac{\partial L}{\partial s_k}\frac{\partial s_i}{\partial w_i} = \sum_{j=1}^{l}\sum_{k=1}^{j} \frac{\partial L_i}{\partial h_1}\frac{\partial h_1}{\partial s_1}\left(\prod_{j=k+1}^{l} \frac{\partial s_j}{\partial s_{j-1}}\right)\frac{\partial s_k}{\partial w_i}$$

$$= \sum_{j=1}^{l} \operatorname{diag}\left(1 - (s_t)^2\right)\Gamma_t(s_{t-1})^x \tag{4.7}$$

$$\nabla r_h = \frac{\partial L}{\partial r_h} = \sum_{j=1}^{t} \frac{\partial L}{\partial s_t}\frac{\partial s_t}{\partial r_h} = \sum_{j=1}^{t}\sum_{k=1}^{j} \frac{\partial L_t}{\partial h_t}\frac{\partial h_t}{\partial s_t}\left(\prod_{j=k+1}^{t} \frac{\partial s_j}{\partial s_{j-1}}\right)\frac{\partial s_k}{\partial r_h}$$

$$= \sum_{j=1}^{t} \operatorname{diag}\left(1 - (s_t)^2\right)\nabla_t(n_t)^{\mathrm{T}} \tag{4.8}$$

$$\nabla b_s = \frac{\partial L}{\partial b_s} = \sum_{j=1}^{b} \frac{\partial L}{\partial s_t}\frac{\partial s_t}{\partial b_s} = \sum_{j=1}^{l} \operatorname{diag}\left(1 - (s_t)^2\right)\nabla_t \tag{4.9}$$

$$\nabla w_o = \frac{\partial L}{\partial w_o} = \sum_{j=1}^{t} \frac{\partial L_t}{\partial h_t}\frac{\partial h_t}{\partial w_o} = \sum_{j=1}^{t} (\tilde{h}_t - h_t)(h_t)^{\mathrm{T}} \tag{4.10}$$

$$\nabla b_o = \frac{\partial L}{\partial b_o} = \sum_{j=1}^{t} \frac{\partial L_t}{\partial h_t}\frac{\partial h_t}{\partial b_o} = \sum_{j=1}^{t} \tilde{h}_t - h_t \tag{4.11}$$

From the above analysis and calculation, it can be seen that the hidden layer h_t in time step t not only depends on the input vector n_t of the current time step t, but also on the calculated output information of the hidden layer h_t in the previous time step $t-1$. Therefore, the final output of RNN in time step t is related to both the input vector n_t of the current time step t and the calculation information of the hidden layer h_t in the previous time step $t-1$.

Theoretically, no matter how long the task sequence to be completed is, the RNN neural network can make full use of all the information to make accurate prediction. However, it relies on partial derivatives and multiplication of the time step t in Eqs. (4.7) and (4.8). When calculating the gradient of the time step, it is also necessary to consider the impact of all implicit state information k before the current time on the implicit state information at the current time. That is to say, since the difference between t and k becomes further and further apart, the number of this effect iteration raises and the corresponding number of partial derivatives and multiplication operations between hidden states increases. When the partial derivative of the activation function is small or equal to zero, this structure of partial derivatives and multiplication increases over time, which causes the RNN gradient to disappear. On the contrary, when the partial derivative of the activation function is large, this structure of partial derivatives and multiplication increases over time, which leads to a gradient explosion in RNN. Consequently, when the input sequence is long, the vanishing and exploding gradients of RNN significantly decrease the model prediction accuracy. LSTM neural network and GRU neural network can effectively solve these problems due to the ability to handle tasks related to long time series [8].

4.2.2 Long Short-Term Memory Neural Network (LSTM)

Long short-term memory neural network (LSTM) was first proposed by Hochreiter and Schmidhuber [9]. Although it has a similar chain structure to traditional recurrent neural network (RNN), LSTM neural network is composed of three classical gates that maintain and adjust the state s_t of the memory unit and hide the state h_t of the cell. Namely, the forgetting gate f_t, the Input and i_t output gates o_t. On the one hand, the LSTM neural network selectively retains and passes through the gate structure to control the information. On the other hand, the sigmoid neural network in the gate structure is fully connected layer and point-by-point multiplication. The integrated not only solves the dependence between the traditional RNN model and the long-term and short-term time, but also effectively deals with the gradient problem in the traditional RNN training process. The sigmoid layer outputs the gate input vector as a real vector of 0–1. When the output vector is equal to 0, any vector multiplied point by point is a zero vector, namely, no information passes. The output vector is equal to 1 and any vector multiplied point by point with it will not change, namely, all information can pass through.

The basic structure of the LSTM neural network is shown as Fig. 4.5. t represents the time step. n_t is the LSTM input vector and h_{t-1} suggests the hidden cell state in the time step t and $t-1$, respectively. The storage cell state of s_t and s_{t-1} suggest in the time step t and $t-1$, respectively. The LSTM neural network element structure shows that both the memory cell state s_t and the hidden cell state h_t in the time step t of the model are transferred to the next time step.

For an input time series $N = (n_1, n_2, \ldots, n_t)$ with a time step size t, each column vector of $N = (n_1, n_2, \ldots, n_t)$ is input into the LSTM neural network in turn. The output h_t is obtained through the LSTM network model for the input n_t at a certain time as follows:

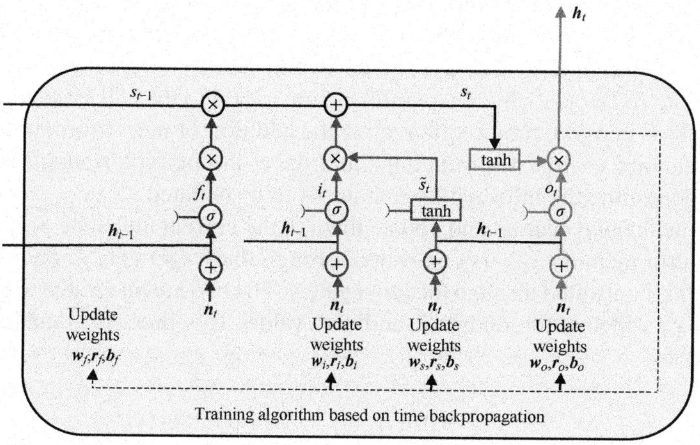

Fig. 4.5 Schematic diagram of the LSTM neural network

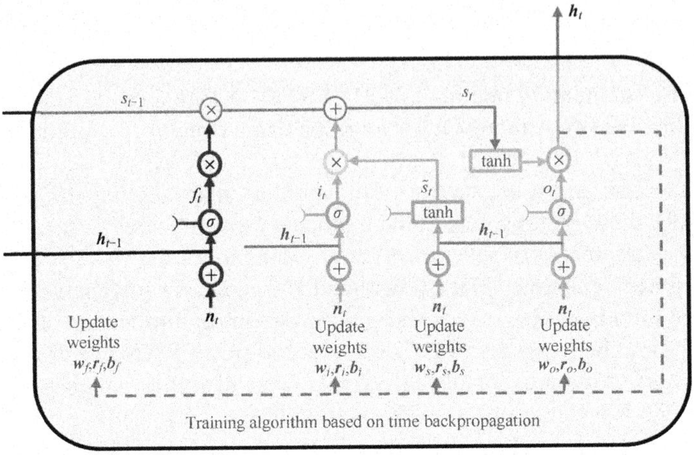

Fig. 4.6 Forget gate principle structure of the LSTM neural network

Forgetting gate f_t: Determine what information from the previous moment cell state s_{t-1} is discarded or retained from the current moment cell state s_t. By reading the output h_{t-1} of the previous moment and the input n_t of the current moment, the gate outputs a vector between 0 and 1 for the cell state at the previous time. "0" means that the information is completely discarded. "1" implies that the information is completely retained. This process is mathematically expressed as:

$$f_t = \sigma\left(w_f(h_{t-1}, n_t) + r_f(h_{t-1}) + b_f\right) \tag{4.12}$$

The trainable w_f, r_f, and b_f represent the connection weight matrix, cyclic weight matrix, and bias matrix of the forget gate f_t of the LSTM neural network, respectively. (h_{t-1}, n_t) is connecting the output h_{t-1} of the previous moment with the input n_t of the current moment. $\sigma(\cdot)$ hints the sigmoid activation function of the LSTM neural network. The principle structure is shown as Fig. 4.6.

Input gate i_t: Decide which new information to add to the cell state at the current moment. Two steps are required to realize the addition of new information:

Generate new vectors $\widetilde{S}_t$ through the *tanh* layer through the *sigmoid* layer in the gate and determine the information that needs to be updated.

Combine the two vectors and update them to the current unit state $\widetilde{S}_t$. In this way, the long-term memory s_{t-1} is controlled through the forget gate f_i, and the current memory $\widetilde{S}_t$ is controlled through the input gate i_t. The two are integrated to form a new unit state s_t, which is the updated candidate value. This process is mathematically expressed as:

$$i_t = \sigma(w_i(h_{t-1}, n_t) + r_i(h_{t-1}) + b_i) \tag{4.13}$$

$$\tilde{s}_t = \tanh(w_s(h_{t-1}, n_t) + r_s(h_{t-1}) + b_s) \tag{4.14}$$

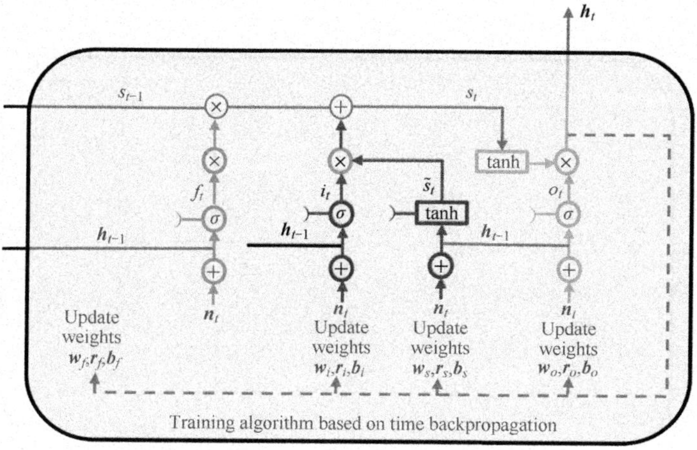

Fig. 4.7 Input gate principle structure diagram of the LSTM neural network

$$s_t = f_t \otimes s_{t-1} + i_t \otimes \tilde{s}_t \tag{4.15}$$

where the trainable w_i, r_i, and b_i indicate the connection weight matrix, the cyclic weight matrix and the bias matrix of the neural network input gate i_t, respectively. $\tanh(\cdot)$ represent the tanh activation function of the neural network. The principle structure is shown as Fig. 4.7.

Output gate o_t: Determine what information in the cell state s_t is output to the current output value o_t of the LSTM neural network. This process is mathematically expressed as:

$$o_t = \sigma(w_o(h_{t-1}, n_t) + r_o(h_{t-1}) + b_o) \tag{4.16}$$

$$h_t = o_t \otimes \tanh(s_t) \tag{4.17}$$

where the trainable w_o, r_o, b_o express the connection weight matrix, the loop weight matrix and the bias matrix of the neural network output gate o_t, respectively. The principle structure is shown as Fig. 4.8.

The LSTM neural network updates all the connection weight matrices and bias matrices of the model by using the time backpropagation training algorithm [10]. In this process, the stochastic gradient descent (SGD) method with loss function (L) is introduced to continuously modify the parameters of the LSTM model to minimize the LF. This process consists of the following three steps:

Step 1: Calculate the output value of each neuron for the LSTM neural network forward through Eqs. 4.12–4.17.

Step 2: Calculate the error term value of each neuron for the LSTM neural network in reverse. The error term Δ_t is inversely transmitted to any moment by

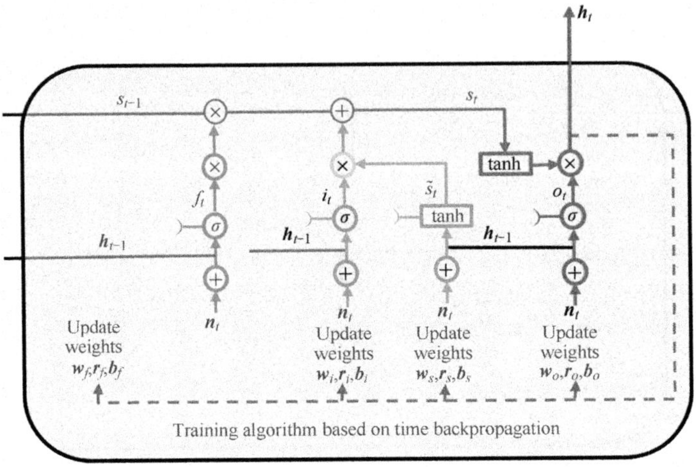

Fig. 4.8 Output gateprinciple structure diagram of the LSTM neural network

the defined error term at the current moment t. The error term Δ_k of the hidden state h_t and the unit state s_t calculated by Eq. (4.17) is transmitted back to any time, as shown in Eq. 4.18. The whole process is expressed mathematically as:

$$\Delta_t = \frac{\partial L}{\partial h_t} \tag{4.18}$$

$$\Delta_{t-1} = \frac{\partial L}{\partial h_{t-1}} = \Delta_t \frac{\partial h_t}{\partial h_{t-1}} \tag{4.19}$$

$$\Delta_k = \prod_{j=k}^{t-1} \Delta_{o,j} w_{\mathrm{oh}} + \Delta_{f,j} w_{fh} + \Delta_{i,j} w_{ih} + \Delta_{\tilde{h},j} w_{\mathrm{ch}} \tag{4.20}$$

$$\frac{\partial L}{\partial \mathrm{net}_t^{l-1}} = \left(\Delta_{f,t} w_{fn} + \Delta_{i,t} w_{in} + \Delta_{h,t} w_{\mathrm{cn}} + \Delta_{o,t} w_{\mathrm{on}}\right) \otimes f'\left(\mathrm{net}_t^{l-1}\right) \tag{4.21}$$

Step 3: Calculate the connection weight matrix gradient, cyclic weight matrix and bias matrix gradient according to the corresponding error terms in Step 2.

The connection weight matrix is divided into two parts: the output vector h_{j-1} at the previous moment and the input vector n_t at the current moment in the backpropagation. As a result, the three gates f_t, i_t, and o_t that the LSTM neural network needs to train and the connection weight matrix with the cell state s_t are divided into two different matrices, namely, w_{fh} and w_{fn}, w_{ih} and w_{in}, w_{sh} and w_{sn}, w_{oh} and w_{on}. The gradient descent of w_{fh}, w_{ih}, w_{sh}, w_{oh} is the sum of the gradient descent at each time, and its mathematical expression is:

$$\nabla w_{\text{fh}} = \sum_{j=1}^{t} \Delta_{f,j} h_{j-1} \tag{4.22}$$

$$\nabla w_{\text{ih}} = \sum_{j=1}^{t} \Delta_{i,j} h_{j-1} \tag{4.23}$$

$$\nabla w_{\text{sh}} = \sum_{j=1}^{t} \Delta_{\tilde{s},j} h_{j-1} \tag{4.24}$$

$$\nabla w_{\text{oh}} = \sum_{j=1}^{t} \Delta_{o,j} h_{j-1} \tag{4.25}$$

The gradient descent of w_{fn}, w_{in}, w_{sn}, and w_{on} is calculated directly from the error term, and its mathematical expression is:

$$\nabla w_{\text{fn}} = \Delta_{f,t} N_t \tag{4.26}$$

$$\nabla w_{\text{in}} = \Delta_{i,t} N_t \tag{4.27}$$

$$\nabla w_{\text{sn}} = \Delta_{\tilde{s},t} N_t \tag{4.28}$$

$$\nabla w_{\text{on}} = \Delta_{o,t} N_t \tag{4.29}$$

The gradient descent of the cyclic weight matrix r_f, r_i, r_s, r_o and the gradient descent of the bias weight matrix b_f, b_i, b_s, b_o are both the sum of the corresponding gradient descent at each moment, which are expressed as mathematically:

$$\nabla r_f = \sum_{j=1}^{t} \Delta_{f,j+1} o_j h(s_j) \tag{4.30}$$

$$\nabla r_i = \sum_{j=1}^{t} \Delta_{i,j+1} o_j h(s_j) \tag{4.31}$$

$$\nabla r_s = \sum_{j=1}^{t} \Delta_{\tilde{s},j+1} o_j h(s_j) \tag{4.32}$$

$$\nabla r_o = \sum_{j=1}^{t} \Delta_{o,j+1} o_j h(s_j) \tag{4.33}$$

$$\nabla b_f = \sum_{j=1}^{t} \Delta_{f,j} \tag{4.34}$$

$$\nabla b_i = \sum_{j=1}^{t} \Delta_{i,j} \tag{4.35}$$

$$\nabla b_s = \sum_{j=1}^{t} \Delta_{\tilde{s},j} \tag{4.36}$$

$$\nabla b_o = \sum_{j=1}^{t} \Delta_{o,j} \tag{4.37}$$

A stable and convergent LSTM neural network model can be established by repeatedly training the LSTM neural network through iterating the above process with specific data. Due to the *sigmoid* function and pointwise multiplication in the input, hidden, and output layers of the LSTM neural network, the gate input vector is output as a real vector of 0–1. Therefore, regardless of the length of the input sequence, the output vector is close to or greater than 1. The gradient of the LSTM neural network will not disappear or explode.

4.2.3 Gated Recurrent Unit Neural Network (GRU)

Although LSTM neural network is widely put to use for its excellent long-term memory function, the training process usually takes a long time due to the complex structure of the neural network. In order to speed up the training process, Kyunghyun et al. [11] proposed a gating unit neural network (GRU) in 2014. As an improved model for the LSTM neural network, GRU has a simpler structure and fewer training parameters while maintaining LSTM performance, which takes less time in the training process.

Compared with the LSTM neural network, GRU neural network only has two gate structures: the reset gate R_t and the update gate U_t. GRU uses the hidden state H_t to replace the cell state s_t in LSTM, transmitting information to the next units traightly. Therefore, GRU neural network does not save or control unit state information. The principle structure is shown as Fig. 4.9.

For an input time series $N = (n_1, n_2, \ldots, n_t)$ with a time step size t, each column vector of $N = (n_1, n_2, \ldots, n_t)$ is input into the GRU neural network in turn. The output h_t is obtained through the GRU neural network model for the input n_t at a certain time as follows:

Reset Gate R_t: Control what information from the previous state H_{t-1} is written to the current candidate state H_t. The smaller R_t implies the less information is written from the previous state to the candidate state. On the contrary, the larger R_t means the

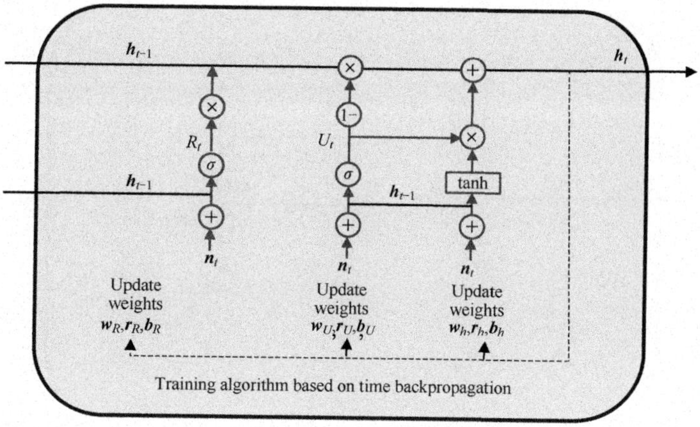

Fig. 4.9 Principle structure diagram of the GRU neural network

more information is written from the previous state to the candidate state. R_t is more effective for short-term sequential tasks and this process is expressed mathematically as:

$$R_t = \sigma(w_R(h_{t-1}, n_t) + r_R(h_{t-1}) + b_R) \tag{4.38}$$

where the trainable w_R, r_R, b_R hint the connection weight matrix, the cyclic weight matrix and the bias weight matrix of the GRU neural network reset gate R_t, respectively, The principle structure is shown as Fig. 4.10.

Update Gate U_t: Determine what information is discarded and what new information is added from the previous moment $t-1$ to the current moment t. The U_t function is similarly to the forgetting gate f_t and input gate i_t in the LSTM neural network but

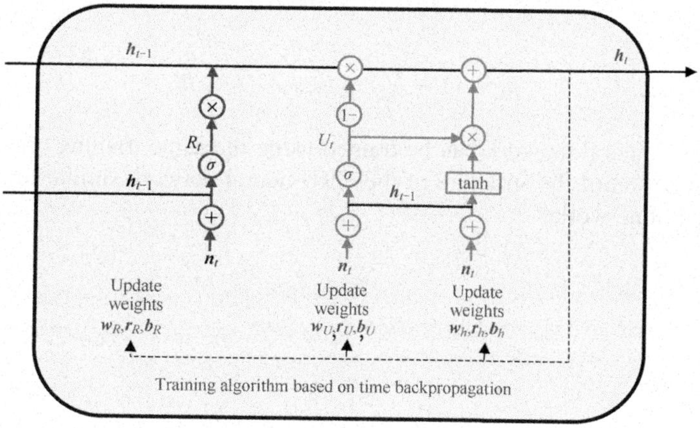

Fig. 4.10 Reset gateschematic diagram of the GRU neural network

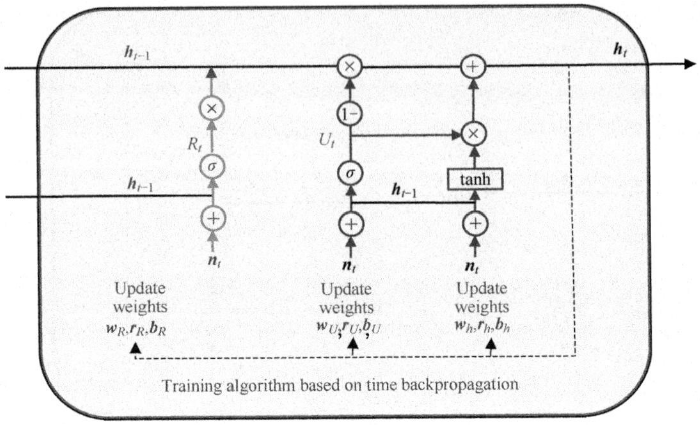

Fig. 4.11 Update gate principle structure diagram of the GRU neural network

it is more effective for long-term sequential tasks. This process is mathematically expressed as:

$$U_t = \sigma(w_U(h_{t-1}, n_t) + r_U(h_{t-1}) + b_U) \tag{4.39}$$

where the trainable w_U, r_U, b_U represent the connection weight matrix, the loop weight matrix and the bias weight matrix of the GRU neural network update gate U_t, respectively, The principle structure is shown as Fig. 4.11.

Different from LSTM neural network, the GRU neural network output vector h_t generates a new vector $\tilde{h}_t$ by the tan$h$ layer, and the vector is directly obtained by combining the vector with the update gate U_t. Namely, the information is directly obtained through the hidden layer h_t, which is mathematically expressed as:

$$\tilde{h}_t = \tanh(w_h n_t + r_h U_t h_{t-1} + b_h) \tag{4.40}$$

$$h_t = (1 - U_t) \otimes h_{t-1} + U_t \otimes \tilde{h}_t \tag{4.41}$$

The GRU neural network can be trained using the same training process as the LSTM by reason of the structure of the GRU neural network similar to that of the LSTM neural network.

4.3 Modeling and Compensation of Orientation Error Based on GRU Deep Learning Neural Network for Bioinspired Polarization Compass

Based on the above analysis, in order to effectively reduce the AoP calculation error caused by the deviation of the zenith point position due to the attitude angle change during the carrier movement process, the GRU deep learning neural network is exploited to model and compensate for the orientation error. The method has been validated through turntable and UAV tests and the results show that the method proposed in this book can effectively improve the heading measurement accuracy for the bioinspired polarization compass. By exploring the mapping relationship between the attitude angle changes of a polarized light compass and heading error, the theoretical content of processing polarized light orientation errors has been further enriched.

In Sects. 4.1 and 4.2, the influence of the attitude angles and the coupling effect between the solar meridian and the carrier axis angle (A-SMBA) on the orientation error for the bioinspired polarization compass has been analyzed in detail. The highly complex nonlinear relationship between them and the respective characteristics of typical machine learning and deep learning neural network have been studied. The GRU deep learning neural network is utilized to construct the relationship between them so as to improve the orientation accuracy for the bioinspired polarization compass in the practical application process. The principle of modeling and compensating the orientation error is shown as Fig. 4.12.

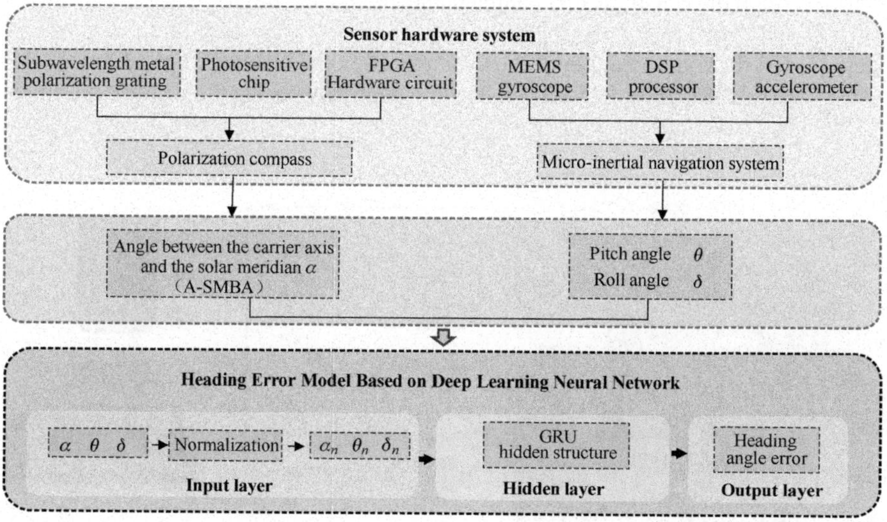

Fig. 4.12 Orientation error model based on the GRU deep learning neural network

In the bioinspired polarization compass orientation error model based on a deep learning neural network, the input vectors are consisted of the angle α, the pitch angle θ and the roll angle δ. α is the angle between the carrier body axis and the solar meridian at the previous moment and provided by the polarization compass. The pitch angle θ and the roll angle δ of the carrier or polarization compass are provided by the micro-inertial navigation system. The heading angle error is calculated by the neural network model and directly output. The high-precision fiber optic gyro inertial navigation system/global satellite navigation system (FOG-INS/GNSS) integrated navigation system (IMU-KVH1750/NovAtel PW7720 in Canada) is used as the measurement reference (reference orientation accuracy of 0.035°). The compass orientation error is obtained by the difference between the measurement reference value and the heading angle output from the bioinspired polarization compass. The heading error of the model output is compared with it to evaluate the performance of the orientation error model proposed in this book.

In order to establish and train the orientation error model, a turntable experiment is first carried out on campus. The turntable test setup is consisted of a self-made bioinspired polarization compass mounted on a three-axis turntable (TBR100), where the turntable test device can continuously generate synchronous model input and output data. The experimental equipment and its characteristics are shown as Fig. 4.13 and Table 4.1 respectively.

The entire test was repeated for 30 days from October 1, 2022 to October 30, 2022 (7:30–11:00 a.m., 15:30–17:00 p.m.). The test process was conducted as follows.

Step 1: Fix the pitch angle θ of the turntable and alter the roll angle δ every 3° from −15° to 15°.

Step 2: Keep the roll angle δ unchanged and change the pitch angle θ every 3° from −15° to 15°.

Step 3: Set the rotation angle of the turntable (i.e., the heading angle ψ) to 0°, 45°, 90°, 135°, 180°, 225°, 270°, 315° respectively during the whole experiment.

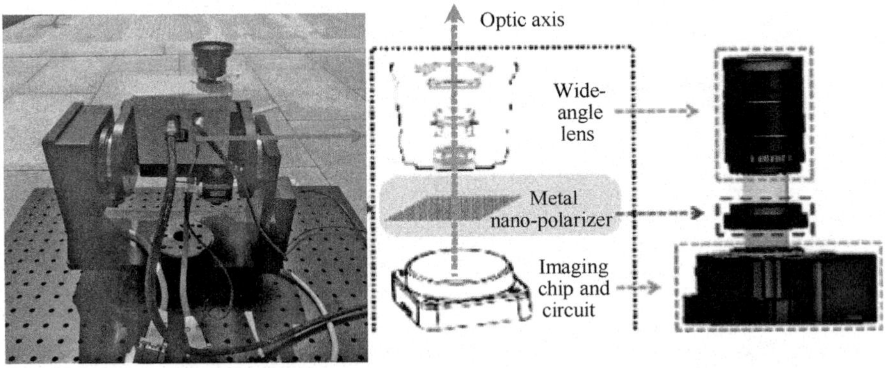

Fig. 4.13 Experimental equipment based on a three-axis turntable

Table 4.1 Characteristics of the rotation experimental equipment

Bioinspired Polarization compass	Polarization unit	High aspect ratio, pixel-scale, sub-wavelength metallic polarization grating
	Photoreceptor chip	purchase
	Hardware circuits	FPGA hardware circuit
Three-axis turntable	Turntable model	TBR100
	Full Step Resolution (°)	0.01°
	Max. speed/(°/s)	20
	Orientation Accuracy (°)	≤0.05

Step 4: Replace the angle between the axis of the carrier body and the solar meridian (A-SMBA) by the angle α between the axis of the carrier body and the solar meridian (PA-SMBA) at the previous moment, which can be obtained in the process of calculating the heading angle.

Step 5: Achieve the different 3D heading error diagrams under the turntable different attitude angles, which are shown as Fig. 4.14.

The results of Fig. 4.14 manifest that the influence of the attitude angles (pitch angle θ, roll angle δ, A-SMBA) from the bioinspired polarization compass on its orientation error is highly complex and nonlinear. The results obtained from the above experimental process are modeled by the GRU deep learning neural network proposed in this chapter. The input of the model is the pitch angle θ, roll angle δ and A-SMBA α. The orientation error is used as the output of the model to establish the relationship between the orientation error and the attitude angle change. The turntable data collected in the above 30 days is used as the model training set, and part of the data obtained in 1 day (460 sampling points) is used as the model test set for validation. The validation results are compared with the following typical

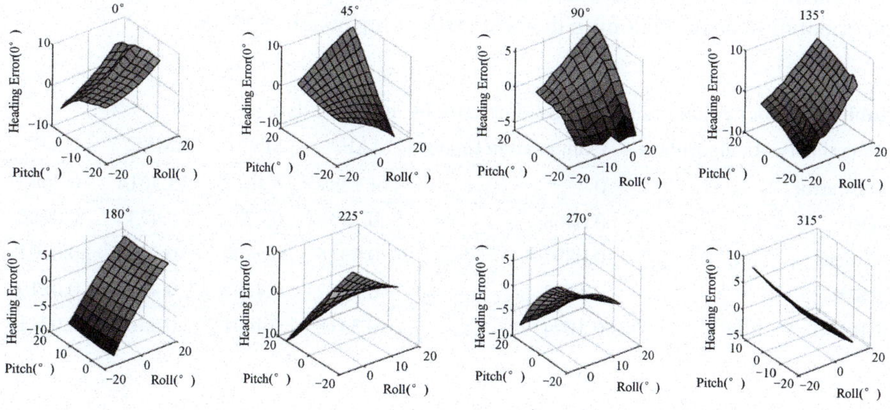

Fig. 4.14 3D Variation of the heading error under different attitude angles

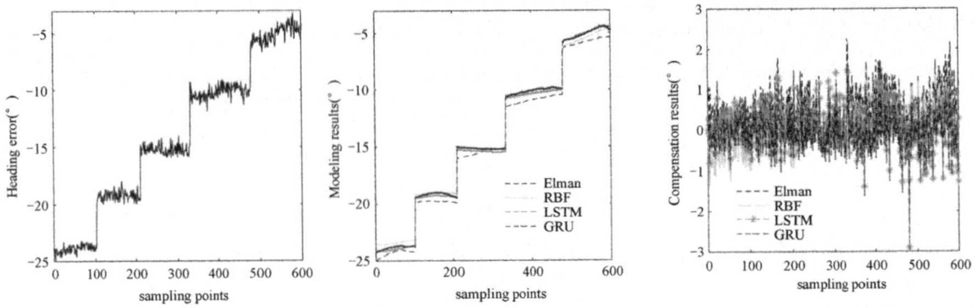

Fig. 4.15 Comparison results of different neural networks

neural networks such as random forest (RF), radial basis function (RBF), Elman neural network (Elman) and LSTM neural network (LSTM). The heading error is compared with the different neural networks mentioned above and are shown as Fig. 4.15 and Table 4.2.

As can be seen from Fig. 4.15, the bioinspired polarization compass orientation error model based on GRU deep learning neural network proposed can effectively reduce its orientation error compared with other typical neural network models. Table 4.2 shows the mean absolute error (MAE), standard deviation (SD), root mean square error (RMSE) of the different neural networks modeling compensation error shown in Fig. 4.15 and the time required for the various neural network models to run the data. From the comparison results with different neural network algorithms in Table 4.2, it can be seen that the smallest root mean square error (RMSE) is 0.4560° obtained after GRU deep learning neural network modeling. Compared with the LSTM neural network, the error model also takes the least time 0.1074 s to run the data, which is mainly due to the fact that the GRU can effectively handle the tasks related to long-time series data by the way of two simplied gates. Therefore, the bioinspired polarization compass orientation error model based on the GRU deep learning neural network can effectively establish a highly complex nonlinear relationship between the attitude angle of the bioinspired polarization compass and its orientation error, obtaining the best error compensation effect.

Table 4.2 Index comparison of the heading error by using test data set

Neural network algorithms	Computation time/s	MAE/°	SD/°	RMSE/°	UA/°
Raw error	0	14.1589	6.4617	15.5614	0.2643
RBF	0.0468	0.3996	0.4702	0.4735	0.0182
Elman	0.0897	0.4315	0.5400	0.5396	0.0191
LS™	0.1173	0.3552	0.4702	0.4735	0.0219
CRANE	0.1074	0.3418	0.4491	0.4560	0.0211

4.4 Experimental Verification of Orientation Error Model Based on GRU Deep Learning Neural Network for Bioinspired Polarization Compass

In order to further verify the effectiveness and practicability of the orientation error model proposed in this book, an UAV test was conducted on campus at sunset (16:10–17:30) on November 12, 2022. The UAV flight test device is composed of the same bioinspired polarization compass as the Sect. 4.4.2 and the measurement reference (UAV model: hexacopter SLM-6S; maximum load: 15 kg; endurance: 20 min). During the airborne test, the UAV flies at an altitude of 310 m to the ground and a flight distance of 500 m. The flight trajectory is shown as Fig. 4.16.

The six-rotor UAV needs to maintain a certain inclination to gain forward and turn power during flight. As shown in the thick curve in Fig. 4.17, the orientation accuracy of the bioinspired polarization compass fixed to the UAV is affected due to the influence of the UAV attitude tilt. The orientation accuracy will be affected more seriously when the UAV turns during flight.

In order to effectively reduce the orientation error for the polarization compass caused by the attitude change during the UAV flight, the orientation error model based on GRU deep learning neural network described in Sect. 4.1.2 is used to compensate. Figure 4.18 shows the orientation error curves after various neural network model compensation. The corresponding orientation error evaluation index results are shown as Table 4.3.

The thick curve displated in Fig. 4.18 represents the polarization compass orientation error compensation result based on the GRU deep learning neural network. As can be seen from Fig. 4.18 and Table 4.3, the result of modeling and compensating the orientation error using the GRU deep learning neural network proposed is 0.5218° (RMSE). The result of orientation error compensation based on RBF

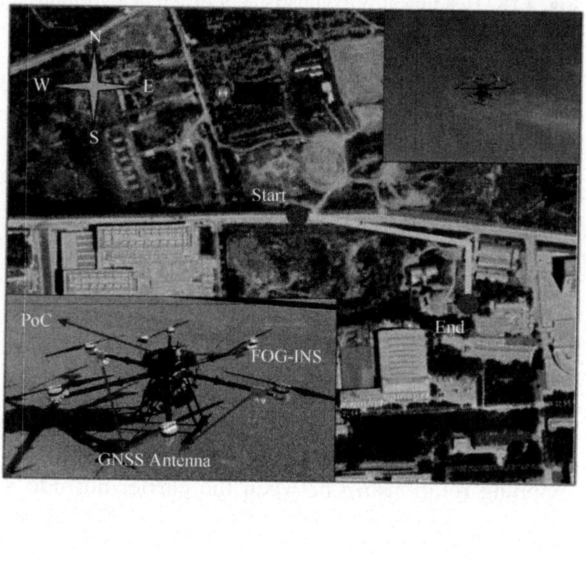

Fig. 4.16 Flight experiment of PC by using the six rotors UAV

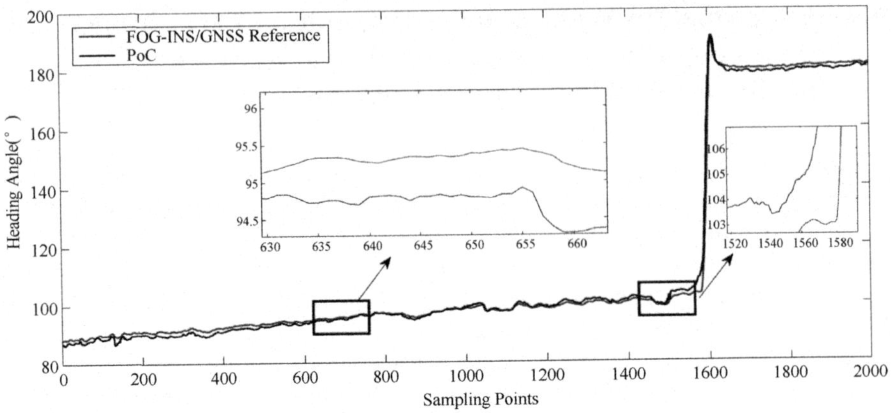

Fig. 4.17 Heading angle comparison between the measurement reference and PC

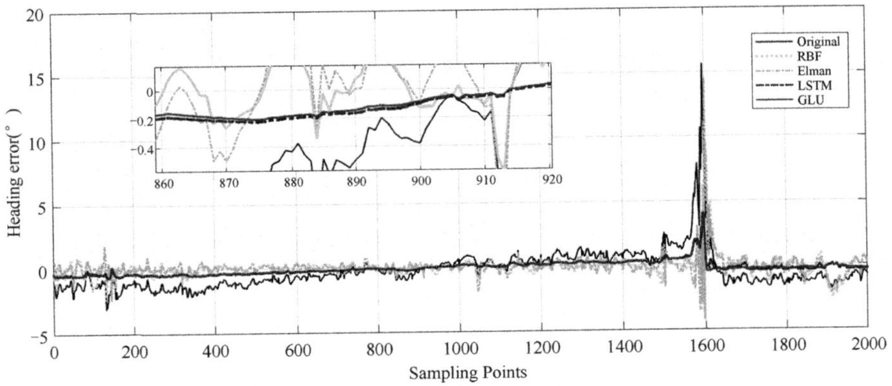

Fig. 4.18 Heading error comparison of different algorithms after compensation

Table 4.3 Heading error indixes after compensating by different neural networks

Neural network algorithms	MAE/°	SD/°	RMSE/°	UA/°
Raw data	0.9597	1.4530	1.4534	0.0115
Elman	0.4341	0.7740	0.7742	0.0118
RBF	0.3950	0.7365	0.7365	0.0173
LS™	0.1069	0.5351	0.5352	0.0181
CRANE	0.1067	0.5217	0.5218	0.0325

and Elman classical machine learning neural network is only 0.7365° (RMSE). The comparison results further illustrate that the orientation error model based on GRU deep learning neural network can not only effectively simulate and approximate the mapping relationship between the carrier attitude angle and the orientation error but also has a robust nonlinear fitting function. The above results indicate that this strategy is able to effectively improve the orientation accuracy for the bioinspired polarization compass in the actual working process.

4.5 Chapter Summary

In this chapter, in view of the bioinspired polarization light compass orientation error caused by the carrier attitude angle change and the coupling between them during the actual carrier movement, the orientation error modeling and compensation method based on deep learning neural network is carried out for the bioinspired polarization compass. Not only does it effectively reduce the polarization compass orientation error, but also it ultimately improves its orientation accuracy significantly.

First, a comprehensive analysis is conducted on the changes in the position of the zenith in AoP images collected by the polarization imaging system. This change causes the error in fitting the solar meridian, leading to the error in AoP calculation and ultimately resulting in orientation error. The coupling effect between the angle A-SMBA and the tilt angles of the carrier (namely pitch angle and roll angle) on the orientation accuracy for a bioinspired polarized light compass is discovered through different dynamic experiments.

Secondly, an orientation error modeling and compensation method is presented for the bioinspired polarization compass based on the GRU deep learning neural network in order to eliminate the influence of the above coupling effect on the orientation accuracy.

Finally, the turntable and UAV test are carried out to verify orientation error the modeling and compensation method based on GRU deep learning neural network for the bioinspired polarization compass. The results demonstrate that the orientation error during the actual carrier movement can be controlled within 0.5218° (RMSE) by using the modeling and compensation method proposed in this book.

References

1. Zhao D, Liu Y, Wu X, Dong H, Wang C, Tang J, Shen C, Liu J. Attitude-Induced error modeling and compensation with GRU networks for the polarization compass during UAV orientation. Measurement. 2022;190:110734.
2. Xiao C, Chen Y. Real-time speech enhancement algorithm based on cyclic neural network. Comput Eng Des. 2021;42(7):1989–94.
3. Sepp H, Jürgen S. Long short-term memory. Neural Comput. 1997;9:1735–80.
4. Shuai G, Yuefei H, et al. Short-term runoff prediction with GRU and LSTM networks without requiring time step optimization during sample generation. J Hydrol. 2020;589:125188.
5. Salinas D, Flunkert V, Gasthaus J. DeepAR: probabilistic forecasting with autoregressive recurrent networks. Int J Forecast. 2017;37(3):1303–4.
6. Zhao D, Wu Y, Wang C, Shen C, Tang J, Liu J. Gray consistency light flow algorithm based on Mask-R-CNN and a spatial filter for velocity calculation. Appl Opt. 2021;60(34):10600–9.

7. Lillicrap TP, Santoro A. Backpropagation through time and the brain. Curr Opin Neurobiol. 2019;55:82–9.
8. Liu Y, Hong Y, Lu Z, Zhang H, Xiong J, Zhao D, Shen C. An optimized pulse coupled neural network image de-noising method for a field-programmable gate array based polarization camera. Rev Sci Instrum. 2021;92(11):113703.
9. Hochreiter S, Schmidhuber J. Long short-term memory. Neural Comput. 1997;9(8):1735–80.
10. Safa OS, Suleyman SK. Non-uniformly Sampled Data Processing Using LSTM Networks. IEEE Trans Neural Netw Learn Syst. 2019;30(5):1452–61.
11. Kyunghyun C, et al. Learning phrase representations using RNN encoder–decoder for statistical machine translation. In: Proceedings of the 2014 conference on empirical methods in natural language processing (EMNLP). Stroudsburg, PA: Association for Computational Linguistics; 2014. pp. 1724–34.

Chapter 5
Seamless Integrated Orientation Method and System for Bioinspired Polarization Compass/Inertial Navigation System

The accuracy of the bioinspired polarization compass (PC) is easily affected by complex environmental conditions such as cloudy weather, tunnels, building occlusion, and so on. It is difficult to provide continuous orientation information, while the error of inertial navigation system (INS) is easy to diverge over time. As a result, the two orientation means above mentioned can be integrated through information fusion methods such as Kalman filter to improve the performance of the entire integrated orientation system with its good autonomy and complementarity with each other. This chapter focuses on the research of a bioinspired polarized light compass/inertial navigation system (PC/INS) seamless integrated orientation method and system based on self-learning multi-frequency residual correction. First, a integrated orientation model for PC and INS is constructed using a multi-frequency volumetric Kalman filter (CKF-MR) fusion algorithm. Secondly, a Cubature Kalman filter (CKF-MRC) fusion algorithm based on multi-frequency residual correction is proposed to address the issues of low data output frequency and overall system orientation accuracy in the PC/INS integrated orientation system when PC information is not affected. Finally, when the PC is temporarily unavailable due to occlusion or other factors, a self-learning seamless integrated orientation method based on long short-term memory (LSTM) neural network is proposed for the PC/INS integrated orientation system. The seamless integrated orientation method based on CKF-MRC proposed in this book has been validated through UAV test. This integrated orientation method is able to effectively improve the data output frequency and orientation accuracy for the PC/INS system when the PC is not affected. When the PC is temporarily unavailable due to occlusion or other factors, it can still maintain high orientation accuracy and ultimately improve the robustness of the entire integrated orientation system.

D. Zhao, *Intelligent Information Processing for Polarization Compass and Inertial Integrated Navigation System*, Navigation: Science and Technology 15,
https://doi.org/10.1007/978-981-97-7135-6_5

5.1 Seamless Integrated Orientation Orientation System for Bioinspired Polarization Compass/Inertial Navigation System

The bioinspired polarization compass/inertial navigation system (PC/INS) seamless integrated orientation system designed in this book is mainly composed of a self-made bioinspired polarization compass (PC), a KY-IMU112 micro-inertial navigation system (MEMS-INS), and a TX2 core board. The physical and internal structure as shown as Fig. 5.1a and b.

PC uses the atmospheric polarization pattern to calculate the heading angle information, and its specific working principles and structure have been described in detail in Chap. 3. The KY-IMU112 micro inertial navigation system and TX2 core board are described in detail as follows. Figure 5.2 shows the hardware architecture of the entire PC/INS seamless integrated orientation system.

(1) KY-IMU112 micro inertial navigation system

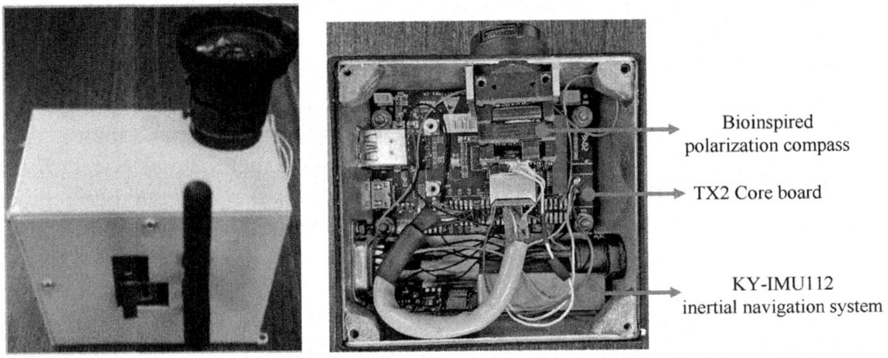

(a) Physical diagram of PC/INS (b) Internal structure diagram of PC/INS

Fig. 5.1 **a** Physical diagram of PC/INS, **b** internal structure diagram of PC/INS

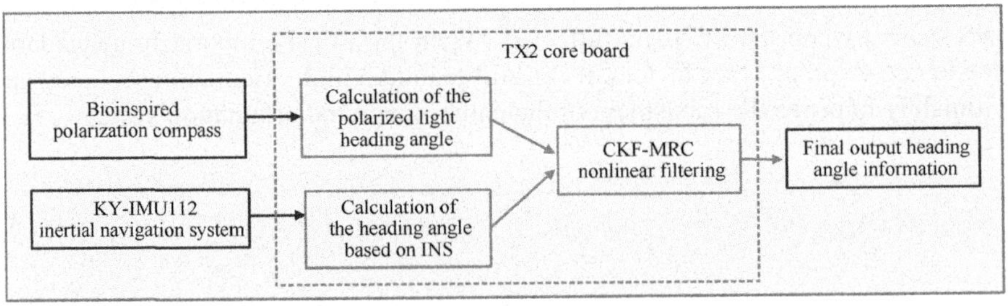

Fig. 5.2 Hardware architecture of the entire PC/INS seamless integrated orientation system

The inertial measurement unit (IMU) in the KY-IMU112 micro inertial navigation system is consisted of three accelerometers and three high-performance MEMS gyroscopes. IMU is responsible for measuring the acceleration and angular velocity of the carrier and sending them to the information processing circuit. The information processing circuit utilizes the acceleration and angular velocity measured by the IMU for navigation data calculation, while receiving satellite navigation information output by the external Global Navigation Satellite System (GNSS) receiver as a reference for integrated navigation and correcting its inertial navigation error. Output navigation information such as pitch, roll, heading, position, speed, and time of the carrier through information interface circuits, ultimately achieving multi-sensor fusion and integrated navigation. When GNSS is invalid, it can output the carrier position, velocity, and heading information calculated by IMU. IMU has a short-term navigation accuracy maintenance function.

(2) TX2 core board

The Jetson TX2 is NIVDIA 256-CDUA-based development board for the AI space, which offers great performance for either single-threaded or multi-threaded processing. With a compact size, 8 GB of RAM and 32 GB of storage, and support for WiFi and Bluetooth, the resources are rich and powerful. In the system designed in this book, the development board needs to solve the heading angle output from the KY-IMU112 inertial navigation system, the polarization light compass and the CKF-MRC nonlinear fusion filter in real time. All things considered, the Jetson TX2 is a good fit for the multi-threaded processing power and small size of the development board required for this book.

5.2 Seamless Integrated Orientation Model Construction for Bioinspired Polarization Compass/Inertial Navigation System

The information fusion inertial-based integrated navigation systems can basically be achieved through the Kalman filter (KF) [1]. In order to improve navigation performance, researchers have proposed a variety of different filtering methods. Chiang et al. [2] proposed an Extended Kalman Filter (EKF) fusion method based on motion constraints and built an error model to obtain reliable measurement results when only INS data was available, using the improved raster-based Simultaneous Localization and Mapping (SLAM) technique to fuse INS data and GNSS data. In Ref. [3], a shipborne landing integrated guidance system was established by fusing visual information, radar, and inertial information by using a joint filter (FF), and the satisfactory performance of the system was verified by on-board tests. In Ref. [4], a Parameter-Centered Difference Kalman Filter (PCD-KF) based on Gaussian process algorithm was proposed, which ultimately improved the prediction performance of Strapdown Inertial Navigation System (SINS) and GNSS data fusion. In order to improve vision and INS performance using PC, Hanxue et al. [5] proposed a robust data fusion method based on the Random Clone Extended Kalman Filter (RC-EKF), which can reduce the long-term growth of vision and INS error.

In addition, with the development of the field of artificial intelligence, artificial neural networks (ANNs), such as backpropagation neural network (BPNN), radial basis function neural network (RBFNN), support vector machines (SVMs), random forests (RF), Elman neural networks (Elman NN), and so on, can be used from the perspective of information processing by accurately constructing an error model through the abstraction of human brain neurons and finally achieving superior prediction performance through the training and prediction phases. Although the above methods have made great improvements in improving the integrated navigation system performance, there are few studies specifically focused on the PC/INS integrated orientation systems. Due to the PC working characteristics, it is necessary to obtain sky polarization information to realize heading measurement. When the PC data is temporarily unavailable under complex environmental conditions (such as cloudy weather, tunnel or building occlusion, etc.), it will lead to the divergence of the data output from the entire PC/INS integrated orientation system. Accordingly, it is of great significance to carry out the research on the PC/INS integrated orientation method and system.

In practical application for the PC/INS integrated directional system, since the fact that the PC sampling frequency being much lower than the INS sampling frequency will result to a decrease in the output frequency and orientation accuracy of the entire integrated directional system data, a CKF-MRC method based on multi-frequency residual correction is proposed to address the problem of in the system when each subsystem is working normally. This method takes the measured heading angle values of PC and INS as the observation variables and state variables of the filter. This fusion algorithm achieves differential optimization fusion for PC data and INS data, which significantly improves the output frequency and orientation accuracy of the entire system. When the PC subsystem is temporarily unavailable, this book exploits a self-learning seamless integrated orientation method based on Long Short-Term Memory (LSTM) neural network to address it. Specifically, LSTM deep learning neural network is adopted to establish a nonlinear relationship among the output value of PC heading angle and the INS z-axis angular frequency and time. By training the LSTM neural network, on one hand, it is able to predict the PC output heading angle data. On the other hand, on the basis of CKF-MRC strategy, it is used to be integrated with INS output heading angle data to enhance the robustness and orientation accuracy of the entire PC/INS integrated orientation system under complex environmental conditions. PC/INS seamless integrated orientation model based on self-learning CKF-MRC is shown as Fig. 5.3.

The above model workflow can be divided into two stages.

Stage 1: Training phase. In the case that the PC subsystem is not affected, the training stage can not only realize the differential fusion of the output data of the PC and INS through the CKF-MRC fusion algorithm, but also construct the complex nonlinear mapping among the z-axis gyroscope angular frequency (ω_z) and time (t) in the INS

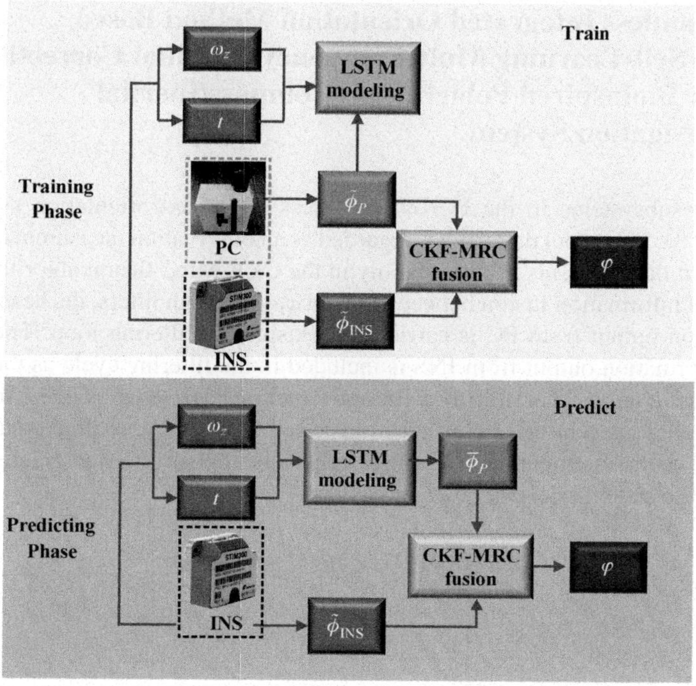

Fig. 5.3 Block diagram of the proposed seamless integrated PC/INS orientation model

and PC heading data ($\tilde{\phi}_P$). The LSTM neural network model takes the INS z-axis angular frequency and time as the model input vector, and the PC heading angle as the model output vector.

Stage 2: Prediction phase. When the PC is temporarily unavailable due to occlusion, the trained LSTM neural network is used to predict the PC heading angle ($\overline{\phi}_P$). The heading data φ is directly output after fusing with the INS output heading data ($\tilde{\phi}_{INS}$) through CKF-MRC. Through the above two phases, the output frequency, orientation accuracy and robustness of the entire PC/INS seamless integrated orientation system are finally effectively improved.

5.3 Seamless Integrated Orientation Method Based on Self-Learning Multi-frequency Residual Correction for Bioinspired Polarization Compass/Inertial Navigation System

When all subsystems in the PC/INS seamless integrated orientation system run normally, the PC output data ($\tilde{\phi}_P$) is regarded as an observation measurement and the INS output data ($\tilde{\phi}_{\text{INS}}$) is a state quantity in the CKF-MRC fusion algorithm. In the process of information fusion between $\tilde{\phi}_P$ and $\tilde{\phi}_{\text{INS}}$ through filters, the heading angle information output from PC is carried out outside the filtering loop. The heading angle information output from INS is included in the filtering cycle as the input of the filter estimation. Therefore, the filtering structure based on the CKF-MRC fusion algorithm designed in this book is an integrated model of a nonlinear state model and a linear measurement model, which forms a nonlinear filtering structure finally. The discrete state model can be expressed as:

$$\boldsymbol{x}_t = f(\boldsymbol{x}_{t-1}, \boldsymbol{u}_{t-1}) + \boldsymbol{w}_{t-1} \tag{5.1}$$

$$\boldsymbol{y}_t = h(\boldsymbol{x}_t) + \boldsymbol{v}_t \tag{5.2}$$

where $\boldsymbol{x}_t$ and $\boldsymbol{x}_{t-1}$ represent the state vectors of the time t and $t-1$ in the PC/INS seamless integrated directional system, respectively. $\boldsymbol{u}_{t-1}$ represents the system filter input vector at the time $t-1$. $\boldsymbol{y}_t$ hints the system measurement vector at time t. $\boldsymbol{w}_{t-1}$ and $\boldsymbol{v}_t$ express the state process and observation process noise at time $t-1$ and t, respectively.

The simplest method of fusing data streams at different frequencies is called a single-frequency Kalman filter, such as the single-frequency Cubature Kalman filter (CKF-SR) [6]. The data fusion process of PC/INS using this filter is shown as Fig. 5.4. T_{PC} and T_{INS} represent the PC and INS sampling intervals, respectively. As can be seen from Fig. 5.4, when the PC output data sample coincides with the INS output data sample, the heading angle data is finally output after the two sample stream data are fused using CKF. Namely, CKF-SR can only be used when the PC output data and the INS output data exist at the same time. Nevertheless, the INS output heading angle data is directly discarded without any processing in the PC sampling interval. Not only does it lead to the accuracy reduction of the whole integrated orientation system, but also the system data output frequency is equal to the lowest PC data output frequency and the real-time performance of the whole system is decreased.

In order to make full use of the high-frequency INS heading angular output data and ensure the real-time performance of the whole PC/INS integrated orientation system, a new multi-frequency Cubature Kalman filter (CKF-MR) is implemented on the basis of SR structure. CKF-MR can be divided into two independent sub-processes: time update and observation update. When the PC output data and INS output data samples coincide, a complete filtering process is performed. A time

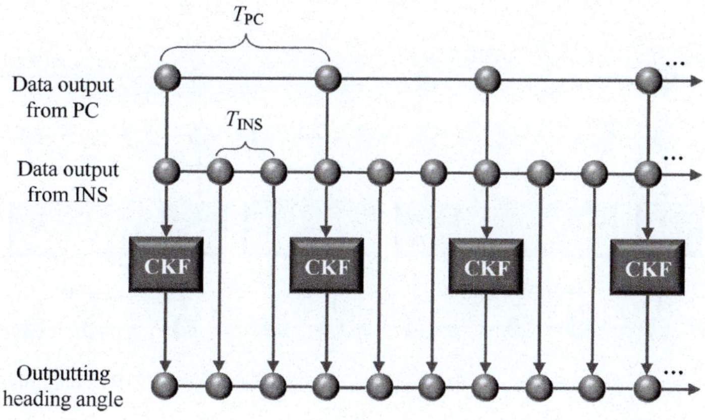

Fig. 5.4 Structure of a single-rate filter for PC/INS data fusion based on CKF-SR

update process is performed and these INS output data samples are used as heading measurements when there are only INS output data samples in the PC sampling interval. The schematic diagram of the PC/INS integrated directional system based on the CKF-MR fusion algorithm proposed in this book is shown as Fig. 5.5.

The heading angle output frequency of the PC/INS integrated orientation system based on the CKF-MR fusion algorithm is the same as the output data frequency of the INS with the highest frequency in the output data stream of the two subsystems. Nevertheless, the CKF-MR fusion accuracy and convergence speed will be greatly affected if the PC is temporarily unavailable under complex environmental conditions and not compensated for the filtering state. Therefore, in order to improve the system orientation accuracy, this book proposes a multi-frequency residual correction Cubature Kalman filter (CKF-MRC) fusion algorithm. The structure of the PC/

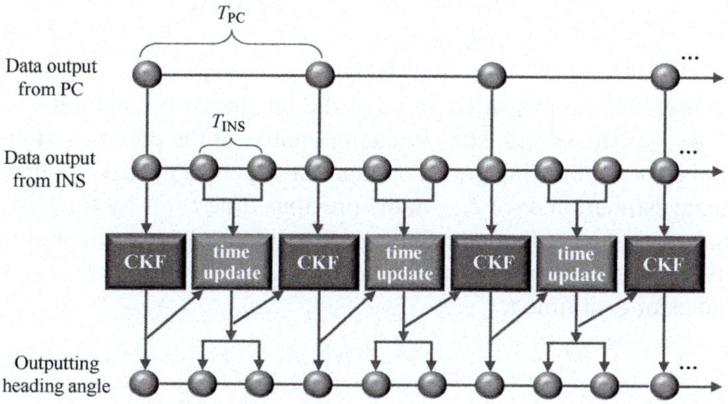

Fig. 5.5 Structure of the multi-rate filter for PC/INS data fusion based on CKF-MR

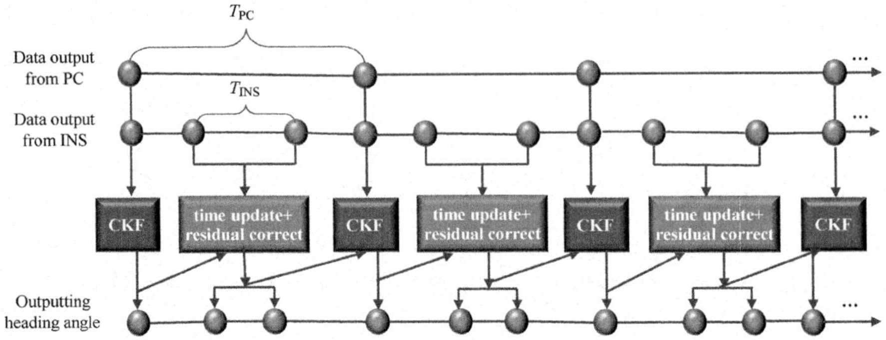

Fig. 5.6 Structure of the multi-rate filter with residual correction for PC/INS data fusion

INS integrated orientation system based on the CKF-MRC fusion algorithm is shown as Fig. 5.6.

The system state estimation error e_t and the residual error ξ_t at the moment t are calculated as follows:

$$e_t = \boldsymbol{x}_t - \hat{\boldsymbol{x}}_t \tag{5.3}$$

$$\xi_t = \boldsymbol{y}_t - H_t \hat{\boldsymbol{x}}_{t|t-1} \tag{5.4}$$

where $\boldsymbol{x}_t$ and $\hat{\boldsymbol{x}}_t$ represent the state value and the optimal estimate at the moment t, respectively. Using Eq. (5.3), the residual error ξ_t of Eq. (5.4) can be rewritten as:

$$\begin{aligned}\xi_t &= \boldsymbol{y}_t - H_t \hat{\boldsymbol{x}}_{t|t-1} \\ &= H_t \boldsymbol{x}_t - H_t \hat{\boldsymbol{x}}_{t|t-1} \\ &= H_t \boldsymbol{\psi}_{t-1} \boldsymbol{x}_{t-1} - H_t \boldsymbol{\psi}_{t-1} \hat{\boldsymbol{x}}_{t-1} \\ &= H_t \boldsymbol{\psi}_{t-1} e_{t-1}\end{aligned} \tag{5.5}$$

where y_t represents the measured value of the integrated orientation system at the moment t. $\boldsymbol{\psi}_{t-1}$ is the system state transition matrix at the previous moment $t-1$. The residual error ξ_t of the integrated orientation system at the time t can be obtained from the state estimation error e_{t-1} at the previous time $t-1$ by Eq. (5.5). Namely, the residual error ξ_t is a function $\xi_t = f(e_{t-1})$ of the state estimation error e_{t-1}. Conversely, the system state estimation error e_t can be represented as $e_t = g(\xi_t)$ by the residual error ξ_t at time t:

$$\begin{aligned}e_t &= x_t - \hat{x}_t \\ &= x_t - \hat{x}_{t|t-1} - k_t\left(y_t - H_t \hat{x}_{t|t-1}\right) \\ &= x_t - \hat{x}_{t|t-1} - k_t \xi_t\end{aligned}$$

$$= \left(H_t^{\mathrm{T}} H_t\right)^{-1} H_t^{\mathrm{T}} \xi_t - k_t \xi_t$$
$$= \left(\left(H_t^{\mathrm{T}} H_t\right)^{-1} H_t^{\mathrm{T}} - k_t\right) \xi_t \tag{5.6}$$

where $\xi_t \approx \boldsymbol{H}_t\left(x_t - \hat{x}_{t|t-1}\right)$, $\left(\boldsymbol{H}_t^{\mathrm{T}} \boldsymbol{H}_t\right)^{-1} \boldsymbol{H}_t^{\mathrm{T}} \xi_t = \left(\boldsymbol{H}_t^{\mathrm{T}} \boldsymbol{H}_t\right)^{-1} \boldsymbol{H}_t^{\mathrm{T}} \boldsymbol{H}_t\left(x_t - \hat{x}_{t|t-1}\right)$, $x_t - \hat{x}_{t|t-1} = \left(\boldsymbol{H}_t^{\mathrm{T}} \boldsymbol{H}_t\right)^{-1} \boldsymbol{H}_t^{\mathrm{T}} \xi_t$. R represents the observed noise that is approximately zero and the matrix $\boldsymbol{H}$ is considered irreversible in Eq. (5.7). Thus, the self-propagating form of the state estimation error e_t at time t can be expressed as:

$$\begin{aligned} e_t &= x_t - \hat{x}_t \\ &= x_t - \hat{x}_{t|t-1} - k_t\left(y_t - \boldsymbol{H}_t \hat{x}_{t|t-1}\right) \\ &= \left(x_t - \hat{x}_{t|t-1}\right) - k_t\left(\boldsymbol{H}_t x_{t|t-1} + R_t - \boldsymbol{H}_t \hat{x}_{t|t-1}\right) \\ &= \left(I - k_t \boldsymbol{H}_t\right)\left(x_t - \hat{x}_{t|t-1}\right) - k_t R_t \\ &= \left(I - k_t \boldsymbol{H}_t\right) \psi_{t-1} e_{t-1} - k_t R_t, \end{aligned} \tag{5.7}$$

When the output data samples of PC and INS coincide, the state estimation error e_t and residual error ξ_t at time t are calculated by Eqs. (5.4) and (5.6), respectively. When only INS output data samples are available, the state estimation error e_t at time t is calculated by Eq. (5.7) using the PC sampling interval. In the PC sampling interval, the covariance matrix satisfies $\boldsymbol{R}_t \rightarrow \infty$ and $k_t \rightarrow 0$, the product can be omitted. Therefore, only the filter parameters in the time update subprocess are relevant and Eq. (5.7) can be approximated as:

$$e_t \approx \psi_{t-1} e_{t-1} \tag{5.8}$$

In the traditional CKF method, the optimal state estimation at the moment t is $\hat{x}_t = \hat{x}_{t|t-1} + k_t \xi_t$. It can be seen that the filter gain is theoretically very small for the PC data output sampling interval. Accordingly, in order to effectively improve the filter gain in the traditional CKF algorithm, the state estimation is adjusted to:

$$\hat{x}_t = \hat{x}_{t|t-1} + \kappa_t \cdot \hat{\xi}_t \tag{5.9}$$

where $\hat{\xi}_t$ is the estimated residual error at the time t. $\kappa_t = \mathrm{diag}\left(\left[\kappa_{1,t}, \kappa_{2,t}, \cdots \kappa_{n,t}\right]\right)$ is used to replace the parameter matrix of filter gain in traditional CKF fusion algorithm so as to determine filter bandwidth and response speed. In order to achieve higher accuracy in fusion, the parameters in $\kappa_t = \mathrm{diag}\left(\left[\kappa_{1,t}, \kappa_{2,t}, \cdots \kappa_{n,t}\right]\right)$ can be set based on experience.

In summary, the PC/INS integrated orientation system based on the CKF-MRC fusion algorithm proposed in this book uses the complete CKF-MRC fusion algorithm to effectively fuse the PC output data $\tilde{\phi}_P$ and INS output data $\tilde{\phi}_{\mathrm{INS}}$ when all subsystems in the PC/INS integrated orientation system are available, in which prepares for the subsequent cycle update of the state estimation error and residual error simultaneously. When the PC output data is not available, Eq. (5.8) is utilized to calculate the state estimation error e_t. Eq. (5.5) is adopted to calculate the residual error ξ_t and Eq. (5.9) is used to update the state estimation. Table 5.1 shows the whole process of PC/INS integrated orientation algorithm based on CKF-MRC fusion algorithm.

When the PC data in the PC/INS integrated orientation system is blocked under complex environmental conditions and becomes unavailable, considering that there is a highly complex nonlinear relationship between the PC output data and the INS output data, the neural network is able to be introduced to establish a mapping model among the PC output heading angle data and the z-axis angular velocity, time information of the INS without accumulating errors over time. It is necessary to consider the time information processing when choosing a neural network. Consequently, a time-related LSTM neural network that can make appropriate trade-offs for historical information according to the system state is introduced in this book. After LSTM is trained and the heading angle data output from PC is predicted, it is effectively fused with the heading angle data output from INS through the CKF-MRC algorithm. The fused results are used as the heading angle data output result of the entire integrated orientation system so as to achieve seamless orientation. Since the specific working principle of the LSTM neural network has been described in detail in Chap. 4, it will not be repeated here. In the training process for LSTM, this book uses the z-axis angular frequency (ω_z) and time (t) of INS as the model input vector, and the PC heading angle ($\overline{\phi}_P$) as the model output vector. Subsequently, the CKF-MRC algorithm is fused with the INS heading angle data ($\tilde{\phi}_{\text{INS}}$) output to obtain the final system output heading angle ($\tilde{\phi}_{\text{INS}}$).

The training dataset is obtained through the UAV test and LSTM is pre-trained in the offline state to ensure that PC can accurately predict the heading information in the state of data interruption. In the experiment, the PC sampling frequency is about 5 Hz, and the INS sampling frequency is 100 Hz, which is consistent with the time update frequency of the CKF-MRC fusion algorithm. Therefore, a total of 9000 valid datasets are used to train LSTM in the 30-min training set acquisition experiment. The UAV orientation range is 0°–360°. Each heading change corresponds to a set

Table 5.1 The whole process of PC/INS integrated orientation algorithm based on CKF-MRC fusion

Algorithm: Multi-frequency residual correction data fusion algorithm
• Judgment: Whether bioinspired polarized compass data $\tilde{\phi}_P$ is available – Available: Use Eq. (5.6) to update the estimation error Use Eq. (5.4) to update the residual error The full CKF data fusion algorithm is used – Not available: Use Eq. (5.8) to update the estimation error Use Eq. (5.5) to update the residual error State variables were estimated using Eq. (5.9)

of INS z-axis angular velocity (ω_z) data and time (t) data collected using high-precision benchmarks. After information collection and storage, LSTM constructs a complex nonlinear mapping relationship between the z-axis angular velocity, time information, and PC heading angle based on INS in an offline state. When the system determines that PC is unavailable, it inputs the INS time information (t) and the z-axis angular velocity (ω_z) into the trained LSTM to predict the PC heading angle ($\overline{\phi}_P$). Immediately, it continues to fuse with the heading angle data output from INS using CKF-MRC to achieve seamless integrated orientation, which is able to improve the robustness and orientation accuracy of the entire PC/INS integrated orientation system ultimately.

5.4 Experimental Verification of the Seamless Integrated Orientation Method for Bioinspired Polarization Compass/Inertial Navigation System

In order to verify the PC/INS seamless integrated orientation method and the system performance under various environmental conditions based on CKF-MRC fusion algorithm proposed in this book, an UAV orientation experiment is carried out on the campus of North University of China. The UAV flight test device is consisted of the same as Sect. 4.3 UAV test setup. The INS sampling frequency is set to 100 Hz. The PC data output frequency is about 5 Hz, which it fluctuates within the range of 4–6 Hz due to the influence of lighting environment. The absolute timescale generated by the timer is used to calibrate the PC and INS output data samples and the installation error was calibrated by the theodolite. The model prediction in the experiment is completed offline training according to the method in Sect. 5.2.

First, the UAV test under the PC data availability for the PC/INS integrated orientation system is selected under sunny weather conditions. It takes about 10 min to complete a round of sunny weather test. The UAV flight trajectory and test results are shown as Figs. 5.7 and 5.8.

The results of the CKF-MRC fusion algorithm proposed in this book are compared with the traditional single-frequency Extended Kalman Filter (EKF-SR), the multi-frequency Extended Kalman Filter (EKF-MR), the multi-frequency residual correction based Extended Kalman Filter (EKF-MRC), the traditional single-frequency Cubature Kalman Filter (CKF-SR), and the multi-frequency Cubature Kalman Filter (CKF-MR). The fusion results are verified and compared as Fig. 5.8 and Table 5.2.

As can be seen from Fig. 5.8 and Table 5.2, the EKF-SR fusion algorithm exhibits poor performance due to the lack of nonlinear fitting ability. The accuracy of the entire PC/INS seamless integrated orientation system is 0.21° (RMSE), which is significantly improved compared with other classical data fusion algorithms. In addition, the length of the PC output data in this experiment is 500. The final output data length of the whole PC/INS seamless integrated orientation system obtained by the

Fig. 5.7 UAV test diagram on a sunny day

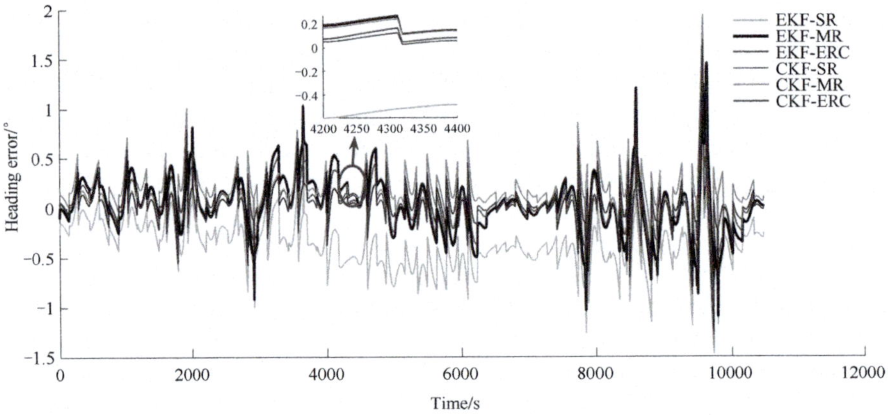

Fig. 5.8 Prediction error of different algorithms for heading data fusion on sunny days

Table 5.2 Comparison of the heading errors of different fusion algorithms on sunny days

Index	EKF-SR	EKF-MR	EKF-MRC	CKF-SR	CKF-MR	CKF-MRC
Variance/°	0.098	0.078	0.051	0.083	0.055	0.043
RMSE/°	0.31	0.28	0.23	0.289	0.23	0.21
UA/°	0.0054	0.0061	0.0075	0.0059	0.0074	0.0066

CKF-MRC fusion algorithm is 10,221, which is significantly higher than that of the pure PC orientation system.

The UAV test is carried out again when the PC/INS seamless integrated orientation system is selected for cloudy weather conditions, namely, the PC is occluded and temporarily unavailable. Due to the limited time when the sun is occluded, the test lasts only 3 min. The UAV flight trajectory and the heading data output of the PC/INS seamless integrated orientation system without LSTM neural network compensation are shown as Figs. 5.9 and 5.10.

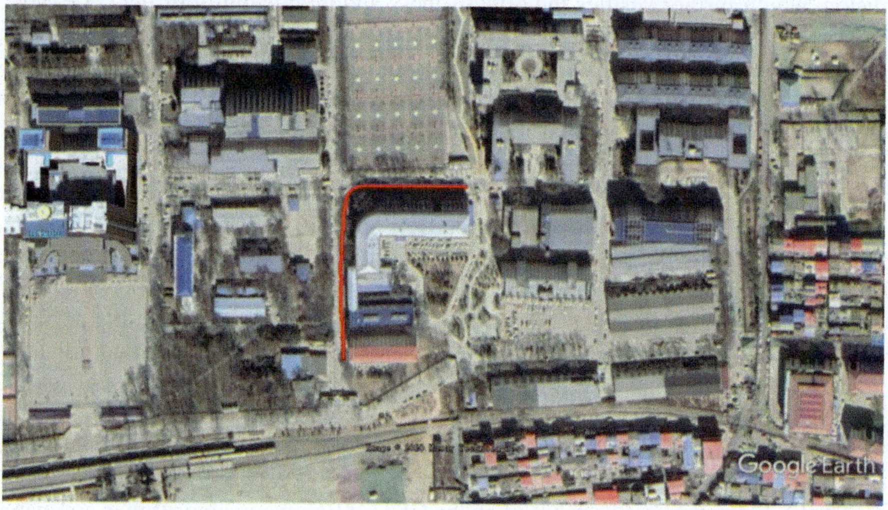

Fig. 5.9 Experiment trajectory of UAV on cloudy days

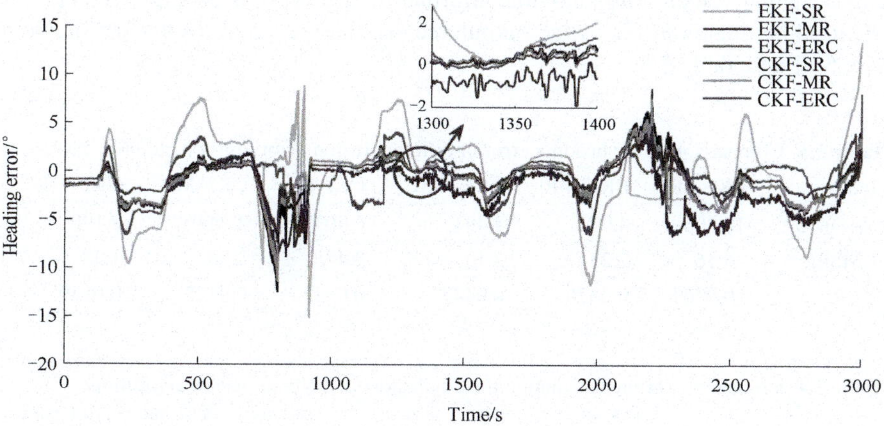

Fig. 5.10 Prediction error of different algorithms for heading data fusion on cloudy days

Under cloudy weather conditions, the CKF-MRC fusion algorithm is verified and compared with the above five classical data fusion algorithms using the same flight test dataset, the results of which are shown in Table 5.3. Compared with the sunny weather conditions, the accuracy of the whole PC/INS seamless integrated orientation system is 1.48° (RMSE) after CKF-MRC processing. The proposed CKF-MRC data fusion algorithm can improve the accuracy by up to 64% compared with other traditional filtering algorithms, which still has obvious advantages. In this experiment, the length of the heading data output from PC is 150. The final output data length of the PC/INS seamless integrated orientation system is 3000 after CKF-MRC fusion, which further verifies the effectiveness and practicability of the CKF-MRC data fusion algorithm proposed.

In addition, the CKF-MRC fusion algorithm is compared with five other classical data fusion algorithms by using a Laptop (Laptop-1D4F19FJ) and a server (Server-5K0HAG3) to compare the heading data output by 10,221 INS data and 500 PC data. The Laptop is equipped with an Inter(R) Core(TM) i7-8550U CPU processor and 8 GB of installed memory and the Server configuration includes an Inter(R) Core(TM) i9-9980XE CPU processor and 256 GB of installed memory, as shown in Table 5.4. The CKF-MRC fusion algorithm proposed runs the heading data through Laptop and Server for 2.884 and 1.782 s, respectively, which is only 0.025 and 0.093 s away from the shortest processing time of EKF-MR. But the accuracy of the whole PC/INS seamless integrated orientation system has been significantly improved. From the above overall experimental results, it can be seen that the PC/INS seamless integrated orientation system based on CKF-MRC fusion has the advantages of best orientation accuracy, highest data output frequency and better data operation efficiency.

The LSTM neural network based on the CKF-MRC fusion algorithm is compared with other classical neural network algorithms such as BPNN and Elman neural network with the same experimental datasets. The LSTM prediction model uses the z-axis angular velocity (ω_z) and time information (t) of INS as the input vectors. The PC heading angle ($\overline{\phi}_P$) is as the output vector. The predicted results are shown as Fig. 5.11 and Table 5.5.

Table 5.3 Comparison of the heading error of different fusion algorithms on cloundy days

Index	EKF-SR	EKF-MR	EKF-MRC	CKF-SR	CKF-MR	CKF-MRC
Variance/°	17.306	4.973	4.494	6.003	4.244	2.190
RMSE/°	4.16	2.23	2.12	2.45	2.06	1.48
UA/°	0.0269	0.0376	0.0447	0.0387	0.0425	0.0759

Table 5.4 Comparison of the calculation time indicators of different fusion algorithms

		EKF-SR	EKF-MR	EKF-MRC	CKF-SR	CKF-MR	CKF-MRC
Time elapsed/s	Laptop	3.025	2.862	2.859	2.889	2.957	2.884
	Server	2.090	1.796	1.743	1.778	1.781	1.782

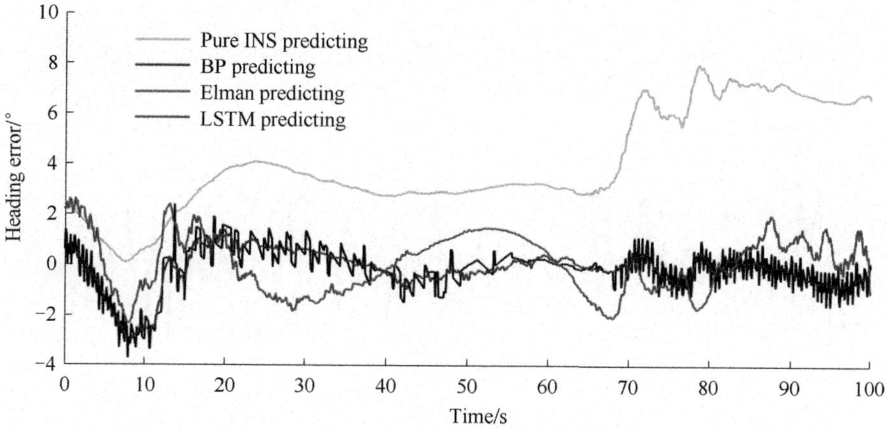

Fig. 5.11 Comparison of predicted heading errors when PC data is not available

Table 5.5 Comparison of predicted heading error indicators when PC data is not available

Index	Pure INS	BPNN	Elman	LSTM
Variance/°	4.56	2.54	1.63	0.93
RMSE/°	2.14	1.59	1.28	0.97
UA/°	0.025	0.036	0.028	0.064

From Fig. 5.11 and Table 5.5, LSTM achieves a smaller heading angle error compared with BPNN and Elman neural network when the PC is temporarily unavailable due to occlusion or other factors. In addition, it can be seen that the predicted performance of BPNN and Elman neural network is inferior to LSTM although they are still better than pure inertial navigation. The orientation accuracy of the latter increases continuously with time drift.

In addition, the heading angle of the PC output is predicted by the LSTM neural network and the predicted value is input into the CKF-MRC fusion algorithm. The results of the heading angle error are shown as Fig. 5.12 and Table 5.6 compared with the other five classical data fusion algorithms.

As can be seen from Fig. 5.12 and Table 5.6, compared with the corresponding CKF-SR fusion algorithm, the EKF-SR and improved fusion algorithms have poor performance due to insufficient nonlinear data fusion capabilities. The CKF-MRC fusion algorithm proposed in this book has the best performance in dealing with nonlinear data fusion compared with the other five classical data fusion algorithms. At the same time, when the PC is temporarily unavailable, the whole PC/INS seamless integrated orientation system can not only maintains high orientation accuracy, but also effectively improves the robustness of the PC/INS integrated orientation system after the LSTM neural network prediction.

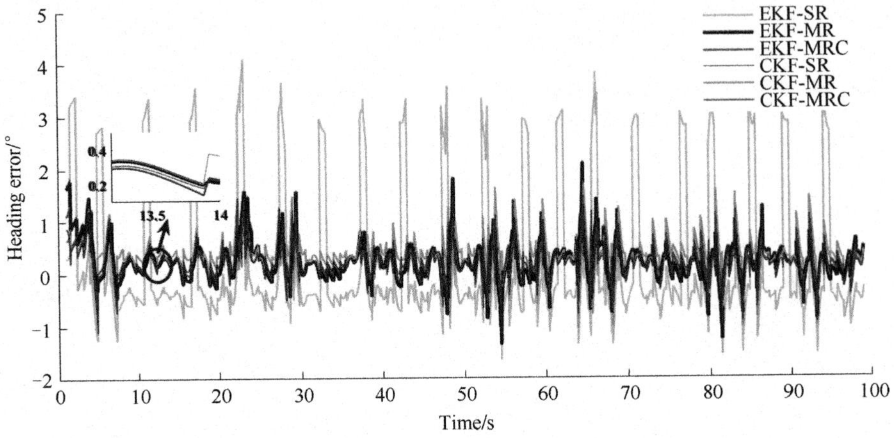

Fig. 5.12 Comparison of the heading error using different fusion algorithms after LSTM

Table 5.6 Comparison of the heading error using different fusion algorithms after LSTM

Index	EKF-SR	EKF-MR	EKF-MRC	CKF-SR	CKF-MR	CKF-MRC
Variance/°	1.29	0.37	0.30	0.37	0.33	0.28
RMSE/°	1.14	0.60	0.55	0.61	0.57	0.53
UA/°	0.0057	0.0066	0.0075	0.0061	0.0074	0.0260

5.5 Chapter Summary

In this chapter, PC and INS are fused into a PC/INS seamless integrated orientation system through a Cubature Kalman Filter (CKF), which is caused by the inconsistent output frequency of the whole system and the temporary unavailability of the PC due to occlusion under complex environmental conditions. The PC/INS seamless integrated orientation method based on CKF-MRC is emphatically studied in this book. This method not only effectively improves the output frequency of the integrated directional system, but also effectively improves the orientation accuracy and robustness of the integrated system under complex environmental conditions.

First, the PC/INS seamless integrated orientation model is constructed by emplying the CKF based on the principle of integrated orientation.

Secondly, when the PC works normally, a Cubature Kalman Filter (CKF-MRC) data fusion algorithm based on multi-frequency residual correction is proposed to solve the problem that the data output frequency of each subsystem in the PC/INS integrated orientation system is inconsistent and the accuracy of the whole orientation system needs to be improved. When PC is temporarily unavailable under the influence of occlusion under complex environmental conditions, a self-learning seamless combinatorial orientation method based on long short-term memory (LSTM) neural network is introduced for the whole PC/INS integrated orientation system.

Finally, UAV various weather conditions experiments are carried out aiming at the PC/INS seamless integrated orientation method based on CKF-MRC fusion proposed in this book. The experimental results demonstrate that the PC/INS seamless integrated orientation method based CKF-MRC can effectively improve the output frequency, accuracy and robustness of the whole integrated orientation system.

References

1. Baolong Y, Donghua Z, Xiaojie L, Xindong W, Deli Y, Chenguang W, Chong S. Inertial/optic flow integrated autonomous velocity measurement based on improved cubature Kalman filtering. Navigat Posit Timing. 2021;9(3):15–9.
2. Chiang KW, Tsai GJ, et al. Seamless navigation and mapping using an INS/GNSS/grid-based SLAM semi-tightly coupled integration scheme. Inform Fusion. 2019;15:181–95.
3. Meng Y, Wang W, et al. A vision/radar/INS integrated guidance method for shipboard landing. IEEE Trans Ind Electr. 2019;11(66):8803–10.
4. Ye W, Li JL, et al. EGP-CDKF for performance improvement of the SINS/GNSS integrated system. IEEE Trans Ind Electr. 2018;2(65):3601–9.
5. Hanxue Z, Chong S, Xuemei C, Huiliang C, Donghua Z, et al. An enhanced fusion strategy for reliable attitude measurement utilizing vision and inertial sensors. Appl Sci Basel. 2019;9(13):2656.
6. Donghua Z, Xiaochen L, Zhao HJ, Chenguang W, Jun T, Jun L, Chong S. Seamless integration of polarization compass and inertial navigation data with a self-learning multi-rate residual correction algorithm. Measurement. 2021;170:108694.

Chapter 6
Summary and Prospect

6.1 Summary of Intelligent Information Processing Technology for Bioinspired Polarization Compass

This book focuses on the issue of autonomous orientation error facing unmanned sports platform in situations such as satellite signal rejection and the accumulated error in inertial navigation system working alone over time. This book provides a detailed introduction to the basic theory and current research status of bioinspired polarized light orientation, the error processing methods of the bioinspired polarized light compasses and the PC/INS seamless integrated orientation method and system. The effectiveness and practicality of the PC/INS integrated orientation error processing method proposed in this book have been verified through various experiments. The main research results and conclusions are summarized as follows.

(1) A multi-scale transformed (MST) denoising method has been presented for the bioinspired polarization compass. First, the mechanism of different types of noise generation for the bioinspired polarization compasses is explored, and their noise characteristics are analyzed. Secondly, in response to the noise in AoP images, a multi-scale transformation stragety based MS-PCA polarization angle image denoising method is proposed by combining the BEMD method and the feature information of the research object. Finally, based on the use of the denoised AoP image for heading angle calculation, a multi-scale time-frequency peak filtering (MS-TFPF) heading angle data denoising method is proposed by combining the EEMD method with the heading angle data noise introduced by the compass circuit. The experimental results illustrate that after denoising the MS-PCA polarization angle image and MS-TFPF heading angle data, the orientation accuracy of the bioinspired polarized light compass is significantly improved (static orientation error RMSE is 0.0735°, turntable orientation error RMSE is 1.2365°, and UAV orientation error RMSE is 0.3116°).

D. Zhao, *Intelligent Information Processing for Polarization Compass and Inertial Integrated Navigation System*, Navigation: Science and Technology 15,
https://doi.org/10.1007/978-981-97-7135-6_6

(2) A novel orientation error modeling and compensation method based on GRU deep learning neural network is proposed for the bioinspired polarization compass. First, a comprehensive analysis is conducted on the factors affecting the orientation error for the bioinspired polarized light compass caused by the carrier attitude angle change (including A-SMBA, pitch angle, and roll angle). Subsequently, combining the cutting-edge achievements of deep learning neural network, a promising orientation error modeling and compensation method based on GRU deep learning neural network is proposed for the bioinspired polarization compass to address the above error. The experimental results prove that the modeling and compensation method proposed in this book can effectively establish a highly complex nonlinear mapping relationship between the carrier attitude angle change and the bioinspired polarization compass orientation error. Ultimately, it can effectively improve the orientation accuracy for the bioinspired polarization compass in the actual carrier application process (turntable orientation error RMSE is 0.4560°, UAV orientation error RMSE is 0.5218°).

(3) PC/INS seamless integrated orientation method based on self-learning multi-frequency residual correction Cubature Kalman Filter (CKF-MRC) is proposed. First, an integrated orientation model is constructed for the PC/INS integrated navigation system using a Cubature Kalman Filter (CKF). Secondly, when the PC operates normally, a data fusion algorithm based on CKF-MRC is proposed to address the issues of inconsistent data output frequencies from various subsystem and the entire orientation system accuracy. Finally, a self-learning seamless integrated orientation method based on long short-term memory (LSTM) neural network is proposed when the PC is temporarily unavailable due to occlusion or other factors. The experimental results show that the seamless integrated orientation method proposed in this book based on CKF-MRC data fusion algorithm can effectively improves the data output frequency and orientation accuracy for the entire PC/INS integrated orientation system.

6.2 Research Outlook

In this book, the bioinspired polarization compass orientation error processing method and the seamless integrated orientation method and system were preliminarily explored and studied, and the phased research results were obtained. However, there are still some issues that need to be studied in depth, mainly including the following aspects.

(1) In the process of integrating a bioinspired polarization compass, a pixel level subwavelength metal polarization grating with high aspect ratio is installed with a photosensitive chip together in the study of this book directly. It is difficult to ensure alignment between the polarization grating pixels and the photosensitive chip pixels during this process. The next step is to explore the preparation method of on-chip integrated subwavelength metal nanograting arrays. The alignment

problem will be solved from a technical perspective and study the integrated error calibration method for the bioinspired polarization compass.

(2) Various algorithms exploited in this book to process the orientation error for the bioinspired polarization compass and PC/INS data fusion are relatively complex and can only be implemented on the upper computer. With the development of artificial intelligence chips, the research on a bioinspired polarization compass based on artificial intelligence chips will be carried out in the future. The complex algorithm transplantation will be implemented to improve the real-time PC/INS orientation accuracy from a system perspective truly. The digital signal processing (DSP) technology is applied to design a new fusion algorithm with lower computational complexity.

The manufacturer's authorised representative in the EU is Springer Nature Customer Service Centre GmbH, Europaplatz 3, 69115 Heidelberg, Germany. If you have any concerns regarding our products, please contact ProductSafety@springernature.com

Printed and bound by CPI Group (UK) Ltd, Croydon, CR0 4YY
01/10/2026
02225919-0005